LIFE
ON MARS

LIFE
ON MARS

THE COMPLETE STORY

Paul Chambers

BLANDFORD

A Blandford Book
First published in the UK 1999 by Blandford,
a Cassell imprint

Cassell plc
Wellington House
125 Strand
London WC2R 0BB

Distributed in the United States by
Sterling Publishing Co Inc., 387 Park Avenue
South, New York, NY 10016-8810

A Cataloguing-in-Publication Data entry for this
title is available from the British Library

ISBN 0 7137 2747 0

Edited and designed by Roger Chesneau

Printed and bound in Great Britain by MPG Books Ltd, Bodmin, Cornwall.

CONTENTS

ILLUSTRATIONS

INTRODUCTION

It is rumoured that in the 1960s a newspaper editor telegrammed an astronomer with the following request: 'Wire one hundred words collect. Is there life on Mars?' The astronomer wired back, 'Nobody knows!', repeated fifty times.[186]

Were the same question asked of the same astronomer today, he or she could justifiably give the same reply. This is not because our knowledge of Mars has not increased since the 1960s—it has, immeasurably—but because the newspaper editor was asking the wrong question. Instead of the predictable 'Is there life on Mars?' question, the editor should have asked 'What is the evidence for life on Mars?', in which case the astronomer would have had trouble giving a reply in less than ten thousand words.

The debate about life on Mars stretches back nearly 400 years and is still very active today. Initially astronomers used their imaginations and religious convictions to guess at what, or who, might be living on the Red Planet. Since then telescopes, Earth-based measuring instruments and, more recently, spacecraft have been used to try to detect any signs of life. Through the centuries many discoveries (and a few mistakes) have been made about Mars, but the fundamental question 'Is there life on Mars?' remains unanswered, although we do now know what type of life could be expected on Mars and where it could live. In 1996 the possible discovery of fossil life in a Martian meteorite rekindled the debate, and with a minimum of ten unmanned spacecraft missions planned to go to Mars before 2005, the issue is set to escalate even further in the coming years.

This book provides the complete story concerning the human search for life on Mars. It charts the narrowing down of our expectations from the first remarks about life in the seventeenth century, through Percival Lowell's observations of canals on Mars, to the modern-day search using interplanetary probes. Current speculation as to where life could exist, or could have existed, on Mars is fully covered, as is the story behind the fossils in the Martian meteorite ALH84001. In recent years there have also been accusations that NASA and others are conspiring to cover up photographic evidence of extraterrestrial civilisations on Mars. These theories, and the history of contact with Martians via mediums, UFOs and other esoteric means, are also covered, as is the role of the science fiction Martian within our psychology.

ACKNOWLEDGEMENTS

I am extremely grateful to the following people, without whose help this manuscript would have been difficult to complete: Dr Mark Biddiss, for his advice and discussions on planetary science and extraterrestrial life; and Peter Hingley, for his help in tracking down obscure references in the Royal Astronomical Society's library. The same thanks also go to the various people who have helped me within the Natural History Museum and within University College, London. I would also like to say a big 'thank you' to Neil Dube, Jim Davy, Phil Dolding, Jon Finch, Colin Clifford, Clare Souter, Dawn Windley, Simon Butler, Karen Pulford and Tim Hackwell for their support and encouragement during the past few years, and also to my nearest and dearest, Rachel Baxter and Sally, Martyn, Matthew and Francis Chambers, for their moral (and financial!) encouragement throughout the years.

PART ONE

THE EARLY
SEARCH FOR LIFE

1. EARLY OBSERVATIONS

THE UNDERSTANDING OF MARS

The quest for life on Mars has been of relatively recent fascination for the inhabitants of Planet Earth. For the majority of recorded history the question of life on other planets has played second fiddle to our desire to understand what the planets are in the first place.

For centuries all our ancestors could see of the planets were points of light in the sky that were brighter and moved differently from the background mass of stars. Mars, with its red hue, was a prominent feature in the night sky and much of the early history of its discovery concentrated on working out its orbit and its relationship to the Sun and to Earth. All thoughts of life were secondary.

We do not know precisely what our ancient ancestors made of Mars, but its striking red colour and prominence in the morning and evening sky would certainly have been noticed. Perhaps they would have created legends around it in the same way that some modern animist religions have. For example, the Dogon tribe in Mali still believes that Mars was formed from congealed blood spilt when its animal god gave birth to the Sun.[225] Indeed, Mars's red colour seems to have inextricably linked the planet with blood and therefore also with violence and warfare. The Babylonians called it the 'Star of Death', to the Ancient Greeks it was the 'Fiery One', and to the Romans it was their god of war as immortalised in Holst's *The Planets*. The Egyptians simply called Mars 'The Red One'—a very apt description.

One factor that drew all the planets to the attention of early astronomers was that, with their circular orbits around the Sun, they appear to wander around the night sky in contrast to the fixed pattern of the stars. Mars's path across the sky was noted for being erratic, as during the course of that planet's two-year cycle the path would appear to double back on itself. Using a model wherein the Earth was at the centre of the universe, with everything else revolving around it, could not help the early astronomers to explain the strange dog's-leg manoeuvre by Mars across the night sky.

This issue defeated the likes of Pliny, Ptolemy and Hipparchus and was left alone as an inexplicable quirk of nature.

The first proposal that the planets and the Earth may revolve around the Sun, rather than the other way round, came from Eudoxus of Cindus in the fourth century BC, but it was not until AD 1543, when this idea was proposed by Nicolaus Copernicus,[50] that the model was taken seriously. However, even with this theory, Mars's unusual movements across the sky could not be fully explained. It was not until 1604 that the problem was finally solved by Johannes Kepler,[113] who was one of the first astronomers to make an intensive study of Mars. He used exact and numerous measurements of the planet to deduce two major facts about its orbit. First, its path is highly elliptical: at its closest point to the Sun it is 206 million kilometres away from it and at its furthest point 249 million kilometres.[115]. Secondly, its axis is slightly inclined to that of Earth, by about 1.8 degrees. This latter fact explained the unusual backward wander in Mars's path across the sky. Kepler was also the first person to calculate the distance between Earth and Mars as well as work out a precise year length for the planet.

In 1609 Galileo Galilei was the first person to turn a telescope towards Mars, but, having a magnifying power of only ×20, he saw next to no detail at all. Better telescopes were developed over the years, and in 1636 Francisco Fontana was the first person to draw a picture of Mars with faint, but discernible, features on it. Subsequent, and improved, drawings were made in 1644 by Father Bartoli and in 1655 by Giambattista Riccoli and Francesco Grimaldi.[68]

The use of the telescope was to bring into play a factor about the orbits of Earth and Mars that was to dominate the observation of the planet from then until the present day. As Mars is further out from the Sun than Earth, it takes 687 (Earth) days to complete a year in comparison to Earth's 365. This means that the two planets only come close to each other every 764 to 810 days. When the two planets are at their closest to each other they are said to be in *opposition*. This is obviously the best time at which to see any surface detail on the planet through a telescope, and the history of the examination of Mars is entirely based around these 25- to 27-month cycles of close approaches. On top of this, every 15 to 17 years the closest approach of Mars to the Sun coincides with its opposition to the Earth. When this happens the opposition distance between the two planets can be reduced to as little as 56 million kilometres compared to a possible 100 million kilometres. Close encounters such as these are called *perihelic oppositions* and are a direct function of Mars's exaggerated oval orbit around the Sun. During perihelic oppositions the surface detail of Mars can be seen

14

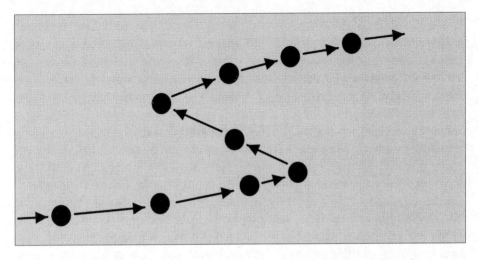

The progression of Mars through the night sky over a period of several months.
The 'dog's leg' in its path is caused by the planet's elliptical orbit.

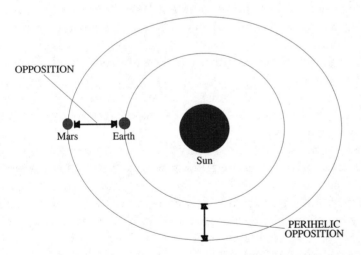

The orbits of Earth and Mars around the Sun, showing how Mars's elliptical
orbit can affect the distance between the two planets.

Fig. 1. Aspects of Mars's elliptical orbit around the Sun.

particularly well, and it was usually in years of very close perihelic opposition, such as 1877 and 1909, that great leaps were made in the study of Mars.

The first reputable drawing of Mars was made in 1659 by Christiaan Hugens, one of Mars's keenest astronomers, who had designed a new type of telescope only five years earlier. Further drawings were made in 1666 by the astronomers Giovanni Domenico Cassini and Robert Hooke. It is from this time that the first speculations about life on Mars began to arise.

THE FIRST THOUGHTS ABOUT LIFE

Although many astronomers had thought about the possibility of life existing on other planets, religious and other social pressures prevented them from recording their comments. The acceptance by the Church of a spherical Earth, a Sun-centred solar system and other astronomical observations may have allowed enough freedom for people to begin wondering what other life might be found beyond Earth. From the mid-seventeenth century onwards all thoughts about life on other worlds were to be known by the term *pluralism*, and plenty of pluralist comments were made in manuscripts over the coming centuries. These were not based on any scientific observations, but on personal expectations about what life on other planets ought to be like. As such, these comments were heavily influenced by the religious, scientific and personal convictions of the commentators concerned. Early pluralist commentators rarely singled out individual planets, such as Mars, for special treatment and dealt instead with a generalised view of life in the solar system or even the universe. Nonetheless, Mars is mentioned a number of times, and these comments give an excellent insight into the thoughts behind pluralism as astronomy developed into a science over the decades.

The earliest reference to the possibility of life existing on Mars was by one of the planet's early astronomical champions, Christiaan Hugens, who remarked to a colleague in 1659 that 'the inhabitants of this last planet [Mars] enjoy days and nights little different from our own.'[68] This thought, and others, was eventually written up and published in a book called *Cosmotheoros*,[101] in which Huygens expanded on his belief that all planets in the solar system are inhabited by flora and fauna similar to those on Earth. He acknowledged that Mars, further from the Sun, would be cooler than Earth, but that we should still expect to find some forms of animal and plant life there, if only on the supposition that if Earth were inhabited then it would be illogical for other worlds be devoid of life. By the time *Cosmotheoros* was eventually published in 1698, four years after Huygens' death, it had already been joined by another early pluralist masterpiece.

In 1686 the book *La Pluralité des Mondes* (The Inhabitation of Worlds) was published by Bernard de Fontenelle, a respected French astronomer. The book follows a popular style of the day and consists of a transcribed dialogue between de Fontenelle and a friend of his, the anonymous Marquis de L——, with whom he spent one summer in France. The debate concerned the evidence for and against the habitation of the planets in the solar system, most of which de Fontenelle believed to contain life. Mars, however, was considered to be uninhabitable and gets scant attention. He merely comments that 'her days are half an hour longer than ours and her years are worth nearly two of our own. It is also five times as small as the Earth and receives much less sun. In short, Mars is not worth the trouble of stopping at. A much prettier choice would be Jupiter with her four moons!'[71]

In countenance to this the Marquis de L—— responds by saying to de Fontenelle, 'A lack of moons is irrelevant! Have you not seen the phosphorus materials which, upon receiving sunlight, soak it up only to release it later. Perhaps Mars has large raised rockfaces of natural phosphorus which could provide extra light at night . . . In America there are birds which radiate enough light to be able to read by [one assumes he is talking about fireflies]. If Mars had a great number of these birds then perhaps, as night fell, they could give light enough for a new day!' It is not recorded by de Fontenelle who actually won this particular debate about Mars, although, with modern hindsight, it was de Fontenelle who was the more correct.

In 1754 Abraham Kastner, a noted poet and anti-pluralist, published a poem about a pluralist friend of his, Christlob Mylius, who had died two years previously. The poem depicts Mylius's soul journeying across the solar system to explore the planets, upon which they debated the possibilities of life. They include Mars, where Kastner has Mylius meeting with the 'eternal souls' of the Martians. Interestingly, a number of mediums in the late nineteenth century were to also report finding dead souls on Mars (see Chapter 14), although none refers to having read Kastner's poem.

Whilst the pluralist debate continued, other astronomers were refining their drawings of Mars and beginning to speculate about conditions on the planet's surface itself. Most notable during this time was William Herschel, who began his study of Mars in 1779. Herschel believed Mars to be very similar to Earth in terms of terrain and climate and produced evidence that the planet had an atmosphere. He may also have been the first to notice a phenomenon that was to play a huge role in the later debate about life on Mars. In 1784 he wrote of seeing 'permanent spots on its [Mars's] surface, and [I] have often noticed occasional changes of partial bright belts'. He may well have been describing the wave of darkening across Mars's surface that would later be attributed to seasonal vegetation growth on the planet. Herschel himself put the darkening down to atmospheric clouds and further commented that 'the people of Mars probably enjoy a situation in many respects similar to ours'. [85]

Other people were now beginning to add their own suggestions about life on Mars. Some of these were pure fantasy, as in Samuel Ellsworth's 1785 almanac, which describes Mercurians as being 'sprightly, small in body, maintaining the upright posture of men, much given to talking, eloquent speech, good lawyers and pettifoggers' whilst the Martians were merely considered to have 'a warlike disposition'. This view of Mars may have been inspired by the Roman association of the planet with the god of war.[63] As well as fanciful speculations, other astronomers tried to apply astronomical theory to the question of plurality. The Compte de Buffon, in 1775, used the size of each planet to estimate a period when it would have been capable of inhabitation. Mars, by his reckoning, would have been cool enough to touch 13,685 years after its formation but would have dried up 56,641 years later, killing all life on it and thus making it barren. Earth, on the other hand, would have cooled later, some 35,983 years after formation, and thus life here still has another 93,291 years to go before the planet dries up.[28] It was commented at the time that 'The genius of Buffon was as much that of a poet as a philosopher.'[65]

Shortly before the end of the eighteenth century another famous poet, William Wordsworth, was to write a poem about a brave adventurer named Peter Bell who travelled the universe visiting planets and meeting their inhabitants, including 'the red-haired race of Mars'.

1830 AND THE SECOND ERA OF MARS

Returning to astronomy, the poor optical quality of eighteenth century telescopes prevented astronomers discerning any firm detail about Mars at all, although some definite features, such as the polar caps and darker surface regions, were mapped. The next phase of Mars observation began in 1830, described by Camille Flammarion in his classic synopsis of the history of Mars as the second era in the history of Mars observation.[68]

In that year the first proper observations of the planet were made through a large purpose-built observatory constructed by Johann Madler and Wilhelm Beer. These astronomers made numerous basic observations about Mars during their lifetimes and realised that the planet's dark markings were not, as previously thought, clouds but actual surface features; they even attempted to name some of them. They could clearly discern lighter and darker regions on the Martian surface and noted that their outline was apt to change between observations. This, they thought, could be due

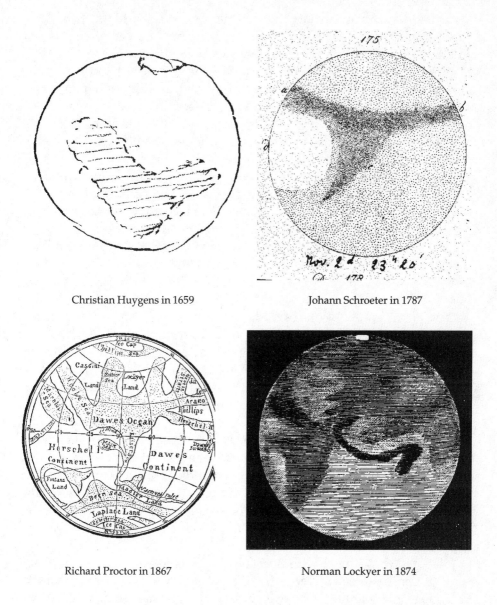

Christian Huygens in 1659   Johann Schroeter in 1787

Richard Proctor in 1867   Norman Lockyer in 1874

Fig. 2. Early drawings of Mars, demonstrating the dramatic improvement in the optical quality of telescopes over 200 years.

to seasonal meltwater from ice caps, making the ground marshy and therefore darker than other, drier regions.[18]

In 1840 Madler and Beer produced the first map of Mars. This, compared to later maps, was merely a sketched drawing containing darker and lighter shadings on the planet's surface. Nonetheless, it contained features that other astronomers could recognise through their telescopes, showing that at least some of these features were fixed landforms on the Martian surface. The recognition that there were discernible and mappable features on Mars re-ignited the pluralist debate about what conditions actually existed on the planet's surface and whether any life could possibly exist there.

At the same time as Beer's and Madler's map, a book was published by Thomas Dick entitled *The Wonders of the Planetary System Displayed: Illustrating the Perfection and Plurality of Worlds*, which contained a census for all the objects in the solar system including the Sun, Moon and rings of Saturn. Mars's population was deemed to be 15,500,000,000 people, which was quite small next to his estimate that the whole solar system contained 21,891,974,404,480 people, most of whom lived on Saturn, Jupiter and Venus.[57]

Dick's book was to be one of the last purely speculative pieces of plurality to be written about Mars and its potential inhabitants. The new era of improved observation was providing actual details from planetary surfaces, especially Mars, and it was these that were to be used in constructing theories about the nature of extraterrestrial life and not mere imagination as before. Although they did not comment specifically on extraterrestrial life, Beer and Madler did start a new line of thinking in the 'life on Mars' debate by suggesting that the darker regions on their map may have been oceans and the lighter regions land. They based this observation on the theory that water would adsorb light and therefore appear darker than land which would reflect it. This was a perfectly valid conclusion for the time and was to conjure up images of an Earth-like planet with oceans, seas, land and rivers and start new thoughts as to what might live on, around and in them.

John Herschel agreed with Beer and Madler, stating that he could see 'with perfect distinctness, the outlines of what may be continents and seas', and suggested that they may even have had a greenish tinge to them. He also believed the polar caps to be made of snow.[84] In the same year Dionysius Lardner, a popular science lecturer, published a volume of his *Popular Lectures* which contained a chapter entitled 'The Plurality of Life'. Part of this included his description of having personally seen clouds on Venus, Mercury and Mars, leading him to conclude that 'whenever the existence of clouds is made manifest, there water must exist; there evaporation must go on; there electricity must reign; there rain must fall; there hail and snow must descend'. It was his belief that God would have populated all the planets in solar system with beings in his own image—in other words, humans could be found on every planet in the solar system, including Mars.

By the beginning of the 1850s the emphasis was shifting away from the old-fashioned speculation about alien civilisations to the creation of an agreed scenario for the existence of life on Mars based on what astronomy was now gleaning from the planet.

2. THE CANALS OF MARS: GIOVANNI SCHIAPARELLI

THE ORIGIN OF THE CANALS

As the Industrial Revolution gathered pace and better techniques were developed for the mass production of high-quality telescopes, so the number of astronomers increased across the world. Not all of these were concerned with studying Mars, but during its oppositions there were few astronomers who did not turn their telescopes toward the planet, even if only for a few minutes. In addition to this growing band of amateurs, universities were setting up professional observatories dedicated to the scientific study of the stars and planets. At one such professional observatory in Rome, the director Angelo Secchi made a discovery that was to have a profound effect on the later debate about life on Mars.

During the opposition of Mars during May 1858 he described seeing a large, blue, triangular patch on Mars's surface. Assuming it to be a sea that separated two continents, he called it the *Atlantic Canale*, meaning 'the Atlantic channel' in the same context as the term English Channel is used on Earth. Unfortunately the Italian word *canale* can also be translated into the word 'canal', implying an artificially constructed waterway, and it was this latter meaning that was to be permanently attached to Mars from then on. Secchi himself was utterly convinced of the existence of seas and continents on Mars, declaring that 'their existence has been conclusively proved'.[214]

Although the issue was far from being conclusively proved, most astronomers of the day agreed with Beer, Madler, Herschel, Secchi and others on the issue of seas and large land masses on Mars. However, in 1862 the Oxford-based geologist John Phillips wondered why, if there were large bodies of water on Mars, the Sun could not be seen reflecting back off them. He thought that maybe the dark regions of Mars were not oceans at all but merely the equivalent of the darker patches on the Moon.[203] This opinion was isolated at the time, and the extent of belief in Mars as an

Earth-like planet can be seen on the most detailed contemporary map of the planet, produced by the Reverend William Dawes in 1864. Another astronomer, Richard Proctor, gave names to the features on Dawes's map that were a clear indication as to what was expected to be found on Mars's surface. Names such as Kaiser Sea, de la Rue Ocean, Secchi Continent, Hind Land and Herschel II Strait reveal an unquestioned belief that the surface of Mars was divided into three main regions consisting of oceans, land and the polar ice caps. The names were never formally adopted as they were considered to be too patronising and inaccurate to be left in place. There was one other interesting feature on Dawes's map, and this, although not commented on at the time, was to be central to Martian studies in the future.

Although the outlines of the continents and seas on Dawes's map are clearly defined, he had drawn in a series of thin dark streaks across the continental regions. Dawes commented that there might be further details hidden within these continents that could not be fully discerned with the telescopes of the day. Within a decade people were to find these details themselves and the thin dark streaks would be interpreted as canals. The year in which this interpretation was made was one of the greatest in the history of Mars's observation.

1877 AND GIOVANNI SCHIAPARELLI'S MAP

By the end of the nineteenth century Mars was becoming a seriously studied planet, with well over a hundred books and papers, as well as several hundred drawings, having been published about it. Unlike opinion at the turn of that century, there was now some general agreement about many of the findings on the planet, including the existence of the darker and lighter regions, clouds and polar ice caps.

In the year 1877 the orbits of Earth and Mars coincided to make the distance between the two planets one of the smallest for nearly half a century. The close proximity of the two planets allowed astronomers to take advantage of recent improvements in telescope design and optical quality, and it is generally agreed that this year marks the start of the modern era in the study of Mars. One astronomer in particular was to spark a renewed enthusiasm in Martian studies amongst his contemporaries and was also, inadvertently, responsible for one of the most embarrassing episodes in the study of Mars.

In 1877 the astronomer Giovanni Virginio Schiaparelli was the director of the observatory in Milan, having already had a highly distinguished career in astronomy. He was particularly noted for his fine eye for detail and technical drawing ability, both of which were used to good effect when he set out to use Mars's perihelic opposition of 1877 to create the finest map yet of the planet's surface.

By making minute measurements of features on the planet's surface, as opposed to the estimation used by his forerunners, he produced a map that was heralded as a masterpiece at the time. Schiaparelli, along with his contemporaries, was unhappy with the names on Dawes's 1864 map and decided to give his map names based on the Bible, Greek mythology and countries. The majority of these names are still in use today, and thus on modern, high-resolution Martian maps we have Libya (Africa) and Ausonia (Italy) separated by the Tyrrhenum Mare (the Tuscan Sea) as well as more esoteric places such as Atlantis, Chaos and Styx. In addition to the continents and seas on his map, Schiaparelli also resolved the wispy dark streaks seen on Dawes's map into a network of definite, long straight lines that, in common with Secchi, he named *canali*—which was to be quickly mis-translated into the English word 'canal'.

Although Schiaparelli was not the first astronomer to observe the *canali*, and not even the first to name them, he was the first person to attach any particular significance to them. Unlike Dawes and Secchi, Schiaparelli was the first to map a profusion of these features and the first to suggest what they might actually represent. He mapped a total of 79 canals and gave each one an individual name, thus formalising their existence as actual physical features on the Martian surface. The canals are marked in as hard, dark lines on Schiaparelli's map and his fascination with them is evident from his description of these features: 'There are, on this planet, traversing the continents, long dark lines which may be designated as *canali*. Those lines run from one to another of the sombre spots that are regarded as seas, and form, over the lighter, or continental, regions, a well-defined network. Their arrangement appears to be invariable and permanent [although] their aspect and degree of visibility are not always the same. They have a breadth of 2 degrees, or 120 kilometres, and several extend for a length of over 80 degrees, or 4,800 kilometres. Every channel terminates at both its extremities in a sea or in another channel.'[203]

Schiaparelli profoundly believed in the existence of the canals but remained neutral on the notion that they might have been artificially produced by neither confirming nor denying the possibility. Instead, he merely conjectured that they represented water-filled channels connecting seas, oceans or large lakes to one another. Schiaparelli published an entire book detailing his 1877 observations on Mars, particularly those concerning the canals. Unlike many publications on Mars, the book received a widespread distribution and subsequently became greatly commented on by other astronomers. An inevitable debate emerged about the validity of the canals with both the 'pro' and 'anti' lobbies becoming deeply entrenched in their beliefs.

DEBATE ABOUT THE CANALS

One of the first people to declare their belief in Schiaparelli's canal-filled map was the keen semi-professional astronomer Cammile Flammarion, who published a paper in 1879 endorsing the prospect of the canals. Flammarion was a great believer in the plurality of life on other planets and by this time had already written books about just such a possibility. He comments that the canals are an absolute certainty and even indulges in a bit of the more old-fashioned speculation about life on Mars by declaring that the lower gravity on Mars would allow Martians to 'fly in its atmosphere'. Soon after this the German esoteric philosopher Jakob Heinrich Schmick also backed Schiaparelli in his unusual book *Der Planet Mars: Eine zweite Erde, nach Schiaparelli*.

The main opposition to Schiaparelli's canals came from a contingent of English astronomers centred around the London-based Royal Astronomical Society. One of the members, Nathaniel Green, had also used the 1877 perihelic opposition to produce his own detailed map of Mars, which differed markedly from Schiaparelli's in both its content and its design. In comparison to the Italian's sharply defined boundaries, dark shading and canals, Green's map was much gentler, with delicate boundaries graduating between lighter and darker shaded regions. Most importantly, Green's map had no canals.

The differences between the two maps can probably be ascribed to the nature of the two astronomers concerned. Schiaparelli was a draughtsman who preferred to measure and ink in every detail in a precise manner using straight, hard lines. Hence his map has very clearly defined dark and light regions with the network of canals running between them. Green, on the other hand, was a renowned artist who had once taught Queen Victoria to paint. He preferred to draw objects in a much more subtle and interpretative manner than Schiaparelli and almost certainly would not have made the numerous minute measurements that the Italian used for his map. The shape and extent of darker and lighter regions on Mars varied between the maps as well. Needless to say, both astronomers believed their map to be the more correct and began publicly to debate the matter.

Green, although sceptical of Schiaparelli's map, was also keen to know why such differences should exist, and he eventually came up with four possible explanations. First, Green wondered whether the fact that Schiaparelli had continued studying Mars for some time after he had finished his mapping would have resulted in some features changing shape in the meantime. Secondly, he thought that many of Schiaparelli's 'hard and sharp lines' were probably due to his tendency to ink boundaries in solidly rather than to graduate them. His third and fourth points were to do with the nature of the canals, which, after all, were the main bone of contention between the two maps.

Green supposed that the canals could be due either to 'a tendency, either in the object glass or the eyepiece [i.e. the telescope], or in the eye of the professor, to elongate and develop dark points seen against a light background' or to atmospheric variations that would make a series of small individual features appear to connect together to form straight lines.[79] These observations were astute and would later be used to explain away the whole phenomenon of the canals on Mars. Despite this, support for Green was thin, and even his announcement that two French astronomers, Etienne Trouvelot and Herbert Sadler, had not been able to see the canals during 1877 did not gain him any further respect.

CLAIMS AND COUNTER-CLAIMS

The debate about the canals of Mars was only just beginning to gather pace amongst astronomers when the next opportunity to view the planet closely came in November 1879. The opposition of Mars that year was a good one, the closest approach occurring during a period of remarkably calm air in Europe, allowing astronomers there excellent views.

After this, Green's limited support began to ebb as more and more astronomers, both amateur and professional, wrote in with their observations of the canals. This list included the world-respected Irish astronomer Charles Burton, who wrote to Green to confirm that he had seen the canals and included a map of Mars with more than a dozen of them on it. Green was a strong admirer of Burton, and after seeing his map of the canals he conceded that 'after Mr Burton's exact description of these forms it would be impossible to doubt their existence'.[79] Other observers included Thomas Webb, Edward Knobel and even the previously sceptical Etienne Trouvelot, who reported seeing nine canals.

Schiaparelli produced a lengthy report on his Martian observations at the Milan Observatory during 1879. Unsurprisingly, these confirmed the presence of the canals, but he also made a number of other observations, including the identification a small white patch in the Tharsis region that he named Nix Olympica. Although he thought that it might have been a snow-covered mountain, we now know this feature to be a giant volcano on the planet's surface. It was also in this report that Schiaparelli first mentions a feature of the canals that was be one of the sensations of that year's opposition of Mars.

Whilst observing the Nilus Canal one night he was shocked to see the single track of the canal resolve itself into 'two tracks regular, uniform in appearance, and exactly parallel'. In other words the canal he was observing was in fact two canals running next door to each other in the same fashion as railway lines. Because of its double nature, Schiaparelli named the feature *gemination*, but it was not be widely commented on until after the next approach of Mars in 1881.

The opposition of 1881 was not a particularly close one but it was nonetheless well observed in the northern hemisphere. Schiaparelli reported finding over 113 canals, including twenty examples of gemination. In his report from that year he commented on the gemination, writing that 'at first I believed this to be the deception of a tired eye, or perhaps the effect of some kind of strabismus [a squint], but I soon convinced myself that the phenomenon was real'.[204] He also noted that the best time for observing gemination was in the months after the closest approach of Mars to Earth and wondered whether it was in fact a seasonal effect of some kind.

In comments made afterwards, Thomas Webb and Flammarion, both passionate believers in life on Mars, quietly endorsed the idea of gemination, as did the Belgian observer Francis Terby. The English astronomers, most notably Green and Edward Maunder, still urged caution over the whole issue of canals and continued to promote their view that they were merely optical illusions.

Maunder commented that although he himself had seen the canals, their position would change from night to night and that significant differences existed between the maps of the canals made by various astronomers. Another observer and previous supporter of the canals, Charles Burton, became converted to Green's and Maunder's sceptical point of view only a few months before his death in 1882, stating that the canals 'are boundaries of differently tinted districts.' [29]

The opposition of 1884 was a very poor one, being neither particularly close nor well suited to observatories in Europe and North America. Publications were still produced about the canals, which, despite pleas for caution from Green and Maunder, were now more or less accepted as genuine by most astronomers. Even the authoritative book *Popular History of Astronomy during the Nineteenth Century* stated that 'the canals of Mars are an actually existent and permanent phenomenon'.[4]

A correspondence arose about the size and quality of telescope that was being used by astronomers to observe the planet. Schiaparelli's 1877 map had been made using an 8-inch reflector, which was small in comparison to Green's 13-inch one. Green made various comments, hinting that the canals might in fact be an optical illusion produced by the glassware in smaller, less optically perfect telescopes. Richard Proctor was later to sum this feeling up when he wrote: 'No one who has ever seen Mars though a good telescope will accept the hard and unnatural configurations depicted by Schiaparelli.'[171]

The 1886 opposition provided ample ammunition against the criticisms concerning the quality of telescopes. Schiaparelli himself had played heavily on the 'life on Mars' issue to get himself funding for a brand new 18-inch refractor telescope, which was operational by 1886. Through it he observed his canals once more, including some which were geminated. Two French astronomers, Henri Perrotin and Louis Thollon, joined the debate in favour of Schiaparelli, claiming to have seen canals, including examples of gemination, through their 15-inch refractor. Both Perrotin and Thollon were to become vocal proponents of the canals, as were other noted figures in the astronomical world including William Denning and Herbert Wilson, both of whom drew maps of the canals in 1886.

The spring and summer of 1888, with its opposition of Mars, were to bring fresh controversy. It started with one of Schiaparelli's supporters, Henri Perrotin, declaring that one of the largest features on Schiaparelli's map, the continent of Libya, had disappeared since the last opposition. He supposed that the sea level must have risen to submerge the entire land mass.

Schiaparelli reluctantly agreed that Libya was disappearing, although he was more cautious about the reason why. Other bad news was on the way for Schiaparelli after a new map, produced using a massive 36-inch refractor at the Lick Observatory by Edward Holden, James Keeler and John Schaeberle, had few features in common with his map or any previous canal maps of Mars. A similar problem was observed with other maps produced in the same year. On this issue Flammarion commented: 'These observations have made us despair. The more one dedicates time, study, and care to the analysis of the numerous and varied observations made of this mysterious planet, the more one is obstructed from arriving at a definite opinion.'[204] He was, nonetheless, still a firm supporter of the canals and advanced the theory that they might be large rivers.

In the meantime Richard Proctor had estimated that to be visible from Earth the canals would have to be a minimum 40 kilometres wide. He thus proposed that perhaps layers of mist above small Martian river valleys could be causing the atmosphere to amplify the size of the valleys to terrestrial astronomers and, in the case of gemination, make them appear to be double.

In 1888, shortly after Proctor's comments were published, William Henry Pickering published a synopsis of the whole canal debate thus far, concluding that they were probably lines of vegetation following rivers on the planet's surface. He also commented that the French astronomer Perrotin had observed a canal running through one of the dark regions which were traditionally thought of as being oceans. He wrote that 'it is clear that either the ocean is not an ocean or the canal is not a canal.'[165] Within a few years Pickering was to become a major player in the canals issue. That year also brought with it an opposition between Earth and Mars, although its poor quality meant that very little was added to the canal controversy.

Similarly, the passing of the 1890 opposition merely brought with it the usual reports from people who had seen the canals and those who had not. By now Green's and Maunder's scepticism was only being supported by a minority of people, although they did manage to get some coverage for their opinion that the canals were probably just soft features made to look solid and straight by over-eager cartographers.

The issue of the canals had reached a stalemate. Few new observations or theories were being added, and, after the initial excitement of their discovery, many astronomers were losing interest in the matter altogether. All this was to change in 1892 when a charismatic businessman named Percival Lowell entered the debate and took the issue of the canals directly to the masses, where their importance would rise to new heights.

3. THE CANALS OF MARS: PERCIVAL LOWELL

CAMILLE FLAMMARION AND WILLIAM PICKERING

The year 1892 brought with it an opposition of Mars, but the major astronomical event of that year was based on Earth not Mars. Camille Flammarion has been mentioned a number of times before in relation to the canals issue. He was an astronomer with a strong belief in extraterrestrial life and had written a number of rather fanciful books on the subject during the late part of the nineteenth century. He had followed the canals issue very closely indeed, and in 1892 he published his book *La planète Mars et ses conditions d'habitabilité*, which was, unlike some of his previous books, to become a classic work on Mars and one that is still widely used today.

This book is still probably the most comprehensive ever written on the history of the study of Mars, and it includes an extraordinary amount of detail regarding people's observations and theories about the planet, especially those to do with the possibility of life existing there. In particular, it has a very good summary of the canals debate and includes literally hundreds of maps and illustrations of the Martian surface.

Although published in French, the book sold very well indeed and widened knowledge about the canals beyond the astronomy cliché of the day. In particular, a copy of *La planète Mars* was to reach the American Percival Lowell, whom we shall discuss shortly. Before talking about Lowell, who was not to enter the canals debate for a couple of years, there are still the results from the excellent 1892 opposition to discuss.

Another highly influential character was to come to prominence that year. William Pickering had only contributed to the Mars debate in a minor way prior to 1892, but in that year his elder brother, the director of Havard University, had sent him and Andrew Douglass to Peru to make measurements of star constellations. However, soon after his arrival Pickering disobeyed his brother and devoted his entire time to observing

Mars. He was soon ordered to return to America, where he proceeded to write four reports based on his Martian observations. Pickering saw canals, including gemination, in profusion, as well as some rather more obscure features, such as snow-capped mountain ranges! Although he had previously commented on the subject, Pickering resolutely stated that some 'canals cross the oceans. If these are really water canals and water oceans, then there would seem to be some incongruity here.'[164] He proposed that perhaps the dark regions were not oceans after all but tracts of vegetation instead. He also noted that at the conjunction of some of the canals were small dark circles, which he named lakes.

Other theories about the canals published in the light of the 1892 opposition suggested that they were an atmospheric phenomenon, that they were geological features and, yet again, that they were river channels. It is interesting to note that no one had publicly proposed the notion that they were of artificial origin. Maunder did his best to promote the dwindling anti-canal cause by suggesting that the whole of Mars would be too cold to support liquid water, but at the end of the day the canal lobby was definitely winning. It was at this point that Percival Lowell takes his place in the history of the search for life on Mars.

PERCIVAL LOWELL

Percival Lowell was born in 1855 to a wealthy and influential Boston family and himself prospered in business. On top of his career, he was also a keen scholar of mathematics, astronomy, geology and classical literature. Other members of his family were similarly distinguished, his sister Amy becoming a famous poet and his brother the president of Harvard for over thirty years. By the 1890s Lowell was very rich and influential and had taken to travelling the world extensively. Although there is some debate over the matter, it is reasonably certain that his obsession with Mars and its canals stemmed from being given a copy of Flammarion's *La planète Mars* as a Christmas present in 1893. He is reputed to have read the whole volume in a matter of days and to have scrawled 'Hurry' across the front page. The urgency behind this statement may have been because the next opposition of Mars was less than a year away, and it is clear that Lowell meant to be able to observe it.

Within a month of receiving *La planète Mars* Lowell had arranged a meeting with the two Pickering brothers at Harvard University. There he proposed setting up an observatory dedicated to planetary studies, especially the study of Mars. Harvard were not keen on the level of involvement that Lowell wanted in the observatory and told him that they preferred their patrons to be of the non-executive variety. However, Lowell was a determined man and he took into his employment the aforementioned William Pickering and his colleague Andrew Douglass, the two disgraced astronomers recalled from Peru, instructing them to find a suitable location to build a modern observatory. Pickering already knew of the perfect place, and within months a modern and well-equipped observatory had been erected on a mountain near Flagstaff, Arizona.

Lowell first observed Mars through his newly built telescope on 24 May 1894 and was from then on to devote a large part of his life to the study of the Red Planet. It was Lowell's belief that the planet should be observed at all possible times, not just during the more favourable periods of opposition. Using this method Lowell, Pickering and Douglass managed to observe Mars at almost every available opportunity over the next few years.

Right from the start, Lowell claimed to be able to see all the canals reproduced on Schiaparelli's map, and he set about making his own maps of the planet's surface. In the period between 1894 and 1916 his observatory was to make 917 sketches of Mars and to find a total of 437 canals crisscrossing its surface, although Lowell personally suspected that there may have been as many 700. He also made many other measurements of the planet and used these to bolster his Martian theories.

By the end of the 1894 opposition Lowell had already put into place much of his theory about the origin and purpose of the canals of Mars. It is clear that most of Lowell's theories are based upon comments and observations made by other astronomers over the years, especially those of William Pickering. In particular, he took Pickering's theories about the dark regions being vegetation, along with his 'lakes' seen at some canal conjunctions and his observations of meltwater flowing away from the polar caps and built them into an all-encompassing theory about the nature of life on Mars. Other academic interests of Lowell's were also woven into the story, especially his fascination with animal evolution, palaeontology and geology, so that by the time he had finished he had done what no other astronomer had dared to until then—compile a full history of Mars from its creation until the current day. The question of life on Mars was central to his history, as were the canals—for which he found a multitude of purposes.

Lowell very publicly promoted his theories on Mars, most notably in a series of lectures which resulted in the publication of three now classic texts entitled *Mars*,[128] *Mars and its Canals*[130] and *Mars as the Abode of Life*.[131] These works were best-sellers around the world and raised an awareness of the planet Mars in the public's imagination that had not hitherto existed. Lowell's lecture tours sold out world-wide and many newspapers of the day gave him wide coverage.

What Lowell promoted was a theory that was bold, consistently argued and also, as we now know, almost totally wrong. The following is an abridged version of Lowell's theory as dictated in his three most popular books.

LOWELLIAN MARS

Lowell's history of life on Mars begins by detailing the origin of the solar system, which he claimed was formed in the aftermath of a massive collision between two giant suns in space. The fiery fragments left behind after this impact gathered themselves into 'balls' of differing sizes which were to become our Sun, its surrounding planets and their surrounding moons. Proof of this, argued Lowell, was to be found in the millions of left-over fragments from the initial collision still floating around in space, such as the asteroids, some of which would fall to Earth as meteorites. All meteorites were therefore the remnants of the formation of the solar system. Whilst we now know this to be partially true, it is also ironic that some have now been proved to be of Martian origin (see Chapter 11).

After a fiery beginning, the next stage in planetary evolution was for each 'ball' to cool down and form a permanent crust. Lowell decreed that the smaller the planet the quicker it would cool and therefore the sooner suitable conditions for life could occur. Size, indeed, is a central feature of Lowell's theory about the evolution of planets and the progress of life in the solar system. It was his idea that because the smaller planets, including Mars and our Moon, would have cooled more quickly and formed a crust earlier than the larger planets, such as Earth, they would therefore have had a head start in their evolution. To Lowell there was a set path along which every planet must travel, which starts with its cooling and finishes with its sterilisation, with a whole set of predestined stages, including the evolution of life, in between. Therefore, the earlier a planet cools, relative to its neighbours, the earlier it will reach the end of this path. This part of his theory shows a remarkable similarity to one proposed by the eighteenth century pluralist the Compte de Buffon (see Chapter 1), who also believed that planets had strict evolutionary paths to follow.

Mars is a small planet in comparison to Earth, and so when Lowell contrasted the two he saw Mars as a planet in its terminal stage before sterilisation whilst Earth, being bigger and therefore younger, was still flourishing. He thus supposed that millions of years previously Mars would have had similar conditions to Earth and that in millions of years' time Earth would look like Mars does now. This can best be summed up by Lowell himself, who wrote that 'It is the planet's size that fits it thus for the role of seer. Its [Mars's] smaller bulk has caused it to age quicker than the Earth, and in consequence it has long since passed through that stage of its planetary career which the Earth at present is experiencing, and has advanced to a further one, to which in time the Earth must come, if it be not overwhelmed beforehand by other catastrophe.'[131]

With regard to the conditions needed for life to exist on modern Mars, Lowell determined that only six elements, carbon, hydrogen, oxygen, nitrogen, phosphorus and sulphur, are required, as well as liquid water, an atmosphere and an equable temperature. By comparing the geology of Earth to that of Mars, Lowell determined that all the necessary elements could be found in the crust of Mars and that water must be present since the planet has ice caps and clouds had been observed by a number of astronomers, including himself and Pickering. Although the planet is further from the Sun than us, the temperature on Mars would be only slightly lower than that of the Earth, said Lowell, because there were so few clouds on the planet to reflect the heat back into space. He predicted that the average global temperature would be 10°C in comparison to the Earth's 18°C. In reality the average temperature is around −40°C, and the planet's thin atmosphere would prevent any heat received by Mars being trapped on the surface. In judging what type of life could exist on Mars, he reasoned that the survival of life could be determined by the highest temperature at which organisms could survive, not the lowest, and cited the fact that many Arizona mountain animals could hibernate through a harsh winter only to awake in the heat of the spring and summer. On Mars the summer was therefore deemed to be the season of life, with winter being the season of hibernation.

Before the building of his observatory Lowell, like his hero Flammarion, had been a firm believer in oceans on Mars. However, in 1894 Pickering had measured for the presence of polarised light reflecting off Mars's oceans and found no evidence of it. This suggested that there were no large bodies of water on the surface. After viewing a number of canals crossing the dark regions, Lowell was converted to Pickering's view that the dark regions must be vegetation growing in the bottom of what were once vast sea beds. Again referring to Mars's evolutionary advancement over the Earth, he hypothesises that the water that had once filled these vast oceans must have escaped, droplet by droplet, into space over a vast period of time, making water a scarce commodity there. Instead of oceans, what moisture was left on the planet now only exists in the polar caps and at the bottom of these once great seas. Plant life was now growing on the ancient sea floors in an effort to trap the last remaining areas of liquid water on the planet. Using the evolution of life as a model, Lowell wrote that the fact that there were such oceans meant that animal life would have more than likely evolved in them, and that as the oceans dried up and shrank, life would have been forced on to the land until, once the oceans dried out altogether, all life on Mars was land-based.

Lowell looked at the Earth and declared that our oceans had been drying up for a number of years. Fossils of marine life found on hills and mountains indicated to Lowell that they were once under water and that the level of the oceans must have dropped dramatically since prehistoric times. In reality, it is normally the land that rises out of the sea with the sea level remaining relatively static, varying by only a few hundred metres over millions of years. He also looked at the belt of deserts that exists to the north and south of the tropics, such as the Sahara, Atacama and Australian deserts, and, with an eye on the prolific deserts of Mars, declared them to be the start of the desertification of Earth. With no oceans, little water and huge deserts, Lowell thought that currently the conditions on Mars were very tough indeed for life and concluded (but without detailing why) that the only creatures capable of surviving such harshness would have to have been intelligent enough to develop their own survival mechanisms. This is odd, considering that the hardiest creatures currently on Earth are not intelligent animals such as monkeys, humans and dolphins but single-celled bacteria (see Chapter 10). Even so, Lowell was convinced that there must be intelligent extraterrestrials on Mars and that, as such, they would have left a mark on the Martian surface capable of being seen from the Earth. This is where the canals come into the story.

As discussed earlier, Lowell spent over twenty years of his life observing and mapping the canals of Mars in detail and came to a number of interesting conclusions regarding their origin, use and dynamics. By the time of his death, Lowell had mapped 437 individual canals on the surface of Mars. He estimated that they varied in length from 450 to 6,200 kilometres and in width from 0.5 to 18 kilometres. All the canals were straight and joined specific areas of the planet, something made possible by Lowell's declaration that the surface of Mars was totally flat, with no mountains or high ground at all. When viewed on a planetary scale, the pattern formed by the canals divided the surface into polygonal areas which Lowell named 'areolas'. He deduced that the areolas must be of recent origin as they were superimposed upon the geology of the planet and were also found in the vegetated ocean basins which, according to Lowell, had only recently dried up.

The dark patches, or 'lakes', that Pickering first observed in 1892 were also found by Lowell, who counted 186 in total. In contrast to Pickering, Lowell named these lakes 'oases'. He also viewed the gemination of ca-

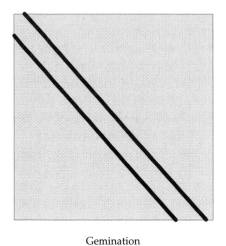

Gemination

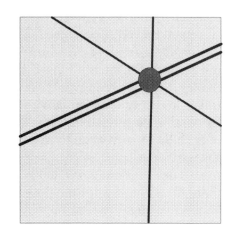

Oases at the conjuntions of canals

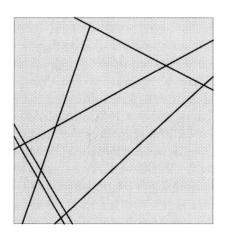

Patterns of canals forming areolas

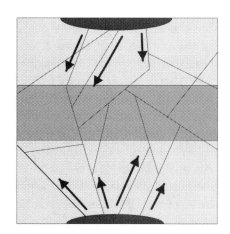

The movement of water from
the poles to the equator

Fig. 3. Aspects of the canals, as seen by Percival Lowell.

nals which was first seen by Schiaparelli in 1879, with 51 examples being personally counted by Lowell. The directional trend of the twinned canals, which was westerly in the northern hemisphere and easterly in the southern hemisphere, fascinated Lowell and led him to conclude that they must have been purposefully built this way in order to utilise the rotation of the planet to move water down them from the planet's poles toward its equator.

In addition to mapping and describing the canal network, Lowell also noted a number of other vital features connected with their observation. In particular, he noticed that the canals were more easily observed at some times of the Martian year than others. He plotted out the visibility of the canals against time and concluded that there was a season on Mars when the canals were much more prominent than at other times of the year. This season approximated to the Martian late spring and early summer, the canals becoming more visible towards the Martian summer solstice and less visible away from it. Lowell also noticed that the canals would become more visible from the poles towards the equator, almost as though they were 'growing' towards the centre of the planet. This was the clue that he needed to help him to reconstruct the nature and function of the Martian waterway network.

The fact that the canals were more visible in the summer than the winter, and that they generally seemed to run from the poles towards the equator, led Lowell to suppose that they had been artificially constructed to take summer meltwater away from the polar ice caps so that it could be delivered to drier parts of the planet. The spread of this summer meltwater down the canal system would stimulate the growth of dormant winter vegetation behind it. Thus it was the seasonal growth of vegetation down the canal network that allowed the canals to be seen better in summer than in winter. By observing the spread of this vegetation on a daily basis, Lowell worked out that it was moving along each canal at the rate of about 92 kilometres day. This equates to a speed of 4 kilometres an hour, which approximates to a slow walking pace! And so it was that Lowell visualised a scenario where each spring a tidal wave of polar meltwater would thunder down the canal network to be followed, a couple of weeks later, by a wave of vegetation growth moving at the same pace.

Vegetation was also used to explain the nature of the oases which only occurred at the conjunction of some of the larger canals. Based on his and Pickering's measurements, Lowell observed the bigger oases having up to seven canals crossing within them. As they did not occur at every canal crossing point, Lowell concluded that they could not be a feature associated with the construction of the canals themselves but, instead, must be the reason for the canals' existence in the first place. As some of the oases were estimated to be 110 kilometres or more in diameter, he declared that they could not be cities in their own right as they would be too big compared to those on the Earth (although not these days). Yet the association of the oases with the conjunction points of the canals suggested that they were the places that the canals were built to irrigate. Lowell therefore concluded that the oases were in fact large patches of vegetation, cultivated by an alien civilisation as a food source, with a normal-sized, Earth-like city at the centre. These cities, being only a few kilometres across, were too small to be seen from Earth; their mantle of vegetation was, however, not, and it was this that astronomers were seeing as oases on Mars.

Lowell's overall vision of Mars was now complete. His theory envisaged a dying planet whose oceans had dried up and whose remaining water was locked away in the polar ice caps. On this planet was a dying race of intelligent beings who had foreseen their own end and were trying to avoid it by using vast canals to transport meltwater from the poles down to the more fertile soil near the equator. The meltwater was used to grow huge fields of edible crops around small city centres, allowing the aliens to survive in what were, otherwise, desert conditions, just as the ancient Egyptians used the flooding of the Nile each year to grow their crops in the Sahara. Lowell's final comment is a sombre one, predicting that the gesture by the Martians is a futile one as Mars is ultimately doomed to dry out altogether, killing its inhabitants. Earth, warned Lowell, will one day end up the same way.

THE REACTION TO LOWELL

Lowell's enthusiasm and talent for self-publicity meant that the canals of Mars received wide coverage amongst the general public and scientists alike, with his books becoming best-sellers and his lecture tours being well attended. In general, though, the public were willing to believe in the possibility of aliens on Mars whereas most astronomers were not. In the public's eye, Lowell, as the owner of a large and modern observatory, was a dedicated scientist whose books were full of mathematical calculations capable of backing up his observations of Mars; to other astronomers and scientists, Lowell was just an enthusiastic amateur who had got hold of the wrong of the stick entirely.

The main complaint against Lowell was that he was too keen to jump to conclusions without having the proper evidence to back himself up. Seasoned astronomer James Keeler complained that Lowell was 'dogmatic and amateurish . . . and draws no line between what he sees and what he infers'.[216] One of his heroes, Schiaparelli, whom Lowell visited in Italy in

1895, commented that 'he needs more experience and must reign in his imagination'.[206] Even one of the astronomers whom Lowell had brought to Flagstaff, Andrew Douglass, once commented that Lowell's methods of research involved 'hunting up a few facts in support of some speculations'.[206] This comment, unsurprisingly, earned Douglass the sack from Flagstaff, although his co-worker Pickering remained loyal throughout.

What is most interesting about the debates on Lowell's theories is that at the turn of the century very few people doubted the existence of the canals or that some form of life would be found on Mars. What they doubted was that the canals were artificial or that life on another planet would be capable of mimicking the achievements of mankind on Earth. There were also doubts about the huge assumptions and jumps in logic that Lowell had used to create his dying alien scenario on Mars in the first place.

Amongst amateur astronomers and the general public it was thoughts about just such a dying civilisation, clinging to life on a desiccated planet, that gave a dramatic edge to the rather cold and clinical world into which astronomy was turning Mars. Lowell's vision of Mars did stimulate many minds, including that of science fiction writer H. G. Wells. The opening lines of his most famous novel, *The War of the Worlds*, were undoubtedly inspired by Lowell and help create an atmosphere of curiosity within which the canals could continue to live in people's minds (see Chapter 15). However, almost as soon as Lowell had built his observatory and got involved in Mars research, the days of the canals as a genuine phenomenon began to be numbered.

4. THE CANALS OF MARS: THE DEATH OF THE CANALS

THE SEEDS OF DOUBT

Percival Lowell built his Flagstaff Observatory in time to measure and observe the close opposition of Mars in 1894. It was from this time that he started to write articles and books specifically to promote his beliefs about the nature of life on Mars. When one looks back at the literature, it is also apparent that, just as Lowell got himself involved in the issue of the canals, many other astronomers were beginning to have grave doubts about the whole affair. The first seeds of this doubt can be seen in the various reports made by astronomers on the 1894 opposition. These would ultimately end in the dismissal of the canals as a genuine phenomenon.

The major publication of that year was by William Campbell, who used a spectroscope to infer that Mars had no water vapour in its atmosphere and was probably uninhabitable for life as we know it.[34] Campbell's results, and the reaction to them, are discussed more fully in Chapter 5, but for now we can say that they contradicted previous measurements and received a generally hostile response. Nonetheless, they did introduce the possibility that Mars may not be the environmentally welcoming planet thus far envisaged and also sparked an era of scientific measurement.

Edward Maunder was still actively working for the anti-canal camp and published a paper on his observations of sunspots into which he managed to introduce the issue of the canals again. He recorded that a series of small sunspots on the solar surface could merge together to form a canal-shaped body. He supposed that at the resolution at which Mars could be observed there would be many smaller features on the surface that would join together to form the canal networks seen by many observers. He cautioned that 'The finest granule, the smallest pore, as we see it, is only the integration of a vast aggregation of details far too delicate for us to detect . . . We have no right to assume, and yet we do habitually assume, that our telescopes reveal to us the ultimate structure of the planet.'[137]

To reinforce this belief, Maunder performed an experiment in which he drew varying sizes of dots and lines placed on a piece of plain paper to see how clearly they could be seen from a distance. He discovered that large dark dots could be distinguished at quite a distance but that smaller dots and lines could not. However, Maunder also discovered that if a series of smaller dots were drawn near to each other they would merge to form a line which could be visible at three times the distance that an individual small dot could be seen.[134] The conclusion of this experiment was, again, that the canals were merely a number of smaller surface features joining together. He finished by quoting that Pickering himself had once 'detected a vast number of small "lakes" in the general structure of the "canal system"', implying that some of the canal observers themselves had unwittingly reported the real cause behind the phenomenon in which they believed.[134] This received little comment, but it was to be the first of a number of similar experiments over the coming decade.

After a series of articles in astronomical journals, Lowell's first book on the canals, simply entitled *Mars*, was published in 1895 and was an instant success amongst the general public. However, as with many books on the paranormal today, just because the public approved it, it did not mean that the science community did as well.

Much of what Lowell had to say about Mars was not a great problem to most astronomers, who for years had proposed and accepted that water, oxygen, vegetation and even the canals could exist on the Martian surface. What they did not like was the theory involving an intelligent civilisation up there building canals and cities in order to stave off their imminent destruction. Reviews of *Mars* were generally split into two groups. The first group, mostly penned by astronomers, thought it a sensational, popular and presumptive piece of work, whilst the second group, mostly written by journalists and amateur astronomers, thought the book a masterpiece and the authority on the current state of Mars research. Amongst the astronomers we have encountered already, those against the book included, predictably, Maunder and Campbell, whilst those in favour, again predictably, included Flammarion, Pickering and Douglass. Others, including Schiaparelli, remained deliberately neutral about it, not wishing to be associated with one side or the other.

In general, it is safe to say that the reaction amongst the astronomical community was generally negative. Lowell's vision of Mars had ignited a fervour amongst the public and within a short period of time there were mediums detailing their telepathic communications with the planet (see Chapter 14) and all manner of discussions about how to communicate with the builders of the canals (see the end of this chapter).

Although the popularity of his book amongst the public was great for Lowell, the astronomical community was frightened by the association of Mars with the esoteric community and many scientists rapidly abandoned their positive or neutral stance on the issue of the canals and began to write articles urging caution or dismissing them altogether. Schiaparelli, in a letter to Francis Terby, summed up the mood of many astronomers at the time perfectly when he wrote: 'Is there a self-respecting man who still risks publicly mentioning this unfortunate planet [Mars] which has become the field of action for all the charlatans of the world; which will . . . supersede the great sea serpent and other similar enticements to the curiosity of the star-crazed?' In other words, Lowell had made the subject of Mars too controversial to tackle by astronomers. In a massive act of irony, Lowell's popularisation of the canals issue was bringing about their demise as a scientifically validated phenomenon.

In the subsequent oppositions of Mars many astronomers avoided the issue of the canals like the plague—except, of course, those who had always been against their existence in the first place. Many still believed that they were there but not that they had been built by an alien race, and as the public was having trouble separating these two issues it was safer not say anything at all. Instead, the astronomical journals contained reports of other measurements and observations taken from Mars. The exception was provided by one Leo Brenner, who published a paper supporting Lowell in every way and detailed the discovery of '68 new canals, 12 seas and 4 bridges' on Mars.[25] Brenner, whose name turned out to be false, was later discovered to be a charlatan and a fraud who made similar exaggerated claims for life on all the planets. He disappeared without trace in 1909.

Lowell remained active at Flagstaff right through the 1896 opposition of Mars and also began observing the planet Venus. In 1896 he published a paper in which he described seeing a network of 'broadish lines on Venus . . . A large number of them, but by no means all, radiate likes spokes from a certain centre.'[129] Although Lowell publicly declared that his 'lines on Venus' were different in nature from those on Mars, he had effectively confirmed Maunder's earlier opinion that the canals were an optical illusion that could be found on every body in the solar system. He lost much support in the astronomical community of which he so desperately wanted to be a part. The events of the previous three years had been too much for Lowell, who suffered a nervous breakdown in 1897 and was not return to his observatory until 1901. In the meantime Pickering and Douglass (until his sacking in 1901) continued to measure and publish material from Flagstaff on his behalf.

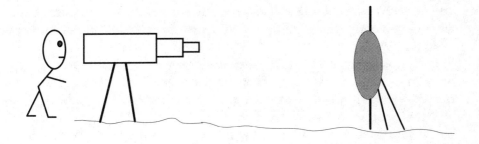

Observing a target through a telescope

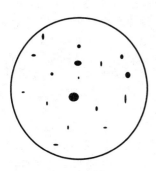

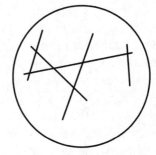

The original target   What people saw through the telescope

Fig. 4. Evans's and Maunder's experiment
to disprove the existence of canals on Mars.

Whereas at one time publications against the canals had only really come from Green and Maunder, they were now being produced by many astronomers keen to distance themselves from Lowell. Eugene Antoniadi believed the canals to be caused by focusing errors in the telescope,[7] whilst many others queued behind Maunder's optical illusion theory. By the time Lowell returned to his observatory to continue his studies of Mars he was facing a still enthusiastic public but an increasingly sceptical astronomical community.

The oppositions of Mars between 1901 and 1907 were generally poor ones and, although the usual measurements and observations were produced about Mars, little was resolved regarding the canals. In fact, probably the most important work on the canals was done here on Earth again by Edward Maunder, who was joined by the astronomer John Evans. In 1903 these two published a paper concerning a series of experiments they had carried out in order to add weight to their optical illusion theory. The experiments involved making a large circular disc with small randomly scattered dots across its surface and persuading a large number of schoolchildren to view the disc through a telescope at varying distances and then to draw what they saw. As they had hoped, at a certain distance from the telescope the children drew not dots, but a network of fine lines that showed a remarkable resemblance to the Martian canals.[138] This was superbly backed up by Captain Percy Molesworth, who, in 1896, had observed the canals perfectly but upon getting a better telescope in 1903 suddenly found that the canals 'are simply the combined results of myriads of [sic] small details, too minute to be appreciated separately.'[150] Other astronomers started to report having had the same experience. Vincenzo Cerulli found that the canal Lethes 'lost its form of a line and altered itself into complex and decipherable system of minute patches',[41] and in 1905 the president of the British Astronomical Society, with regard to Maunder's work, declared that 'The onus of proof now lies upon those who thought the canals were there.'[62]

Lowell was not to be outdone, and in the same year as the British Astronomical Society's statement he released a series of photographs taken at his observatory which purported to show the canals in all their glory. For a short while the canals came back into fashion amongst astronomers, with many, having been silent on the issue for years, again declaring their support for them. The jubilation was short-lived, however, as the same accusations about wishful thinking and small objects joining together to make canal-like shapes were levelled at the photographs, which were discovered only to show canals when greatly reduced. Any attempt at enlarging the photographs resulted in the canals disappearing altogether.

Even so, Lowell's quest had been given a boost, and in 1906 he published his second book on the subject, entitled *Mars and its Canals*.[130] This was little different in style and content from *Mars*, but still sold very well amongst the public. Despite this, the anti-canal lobby had by now won the debate with astronomers, if not the public, as paper after paper was published urging caution about the matter, many lending support to Maunder's and Evans's earlier work. However, the death of the canals was not to be firmly established until 1909.

1909

The autumn of 1909 produced the best perihelic opposition of Mars since the famous one of 1877 which had raised the issue of the canals in the first place. The excellent views that astronomers received of Mars led to a large number of publications, most of which argued against the canals. One astronomer in particular, Eugene Antoniadi, had become acknowledged as the foremost authority on Mars, and of the twenty-odd papers that he produced on his 1909 observations he concluded that 'The "continental regions" of the planet are variegated with innumerable dusky spots of very irregular outline and intensity, whose sporadic groupings give rise, in small telescopes, to the "canal" system of Schiaparelli.'[9] Beyond this, new photographs were being produced of Mars which did not show any evidence of the canals, and as more and more reports of a canal-free Mars came through it became apparent that, after thirty years of fighting, Maunder and his colleagues had won their argument about the nature of the Martian canals.

One of the most public contradictions of Lowell came from Alfred Wallace, Charles Darwin's co-worker, who published a book entitled *Is Mars Habitable?*[239] This work was a direct attack on Lowell's three books and explained in detail why Mars was liable to be a cold, barren planet with no atmosphere. He describes as 'mad' any race of beings who would attempt to build canals across barren deserts under cloudless skies, commenting that any water would not make it for more than a few kilometres before evaporating or soaking away. Percival Lowell, however, refused to accept the collective decision of his beloved astronomical community and continued to promote his dying alien civilisation theory in a series of worldwide lecture tours that were, by and large, well received by the public and resulted in the book *Mars as the Abode for Life*.[131] Lowell died in 1916, still convinced of the existence of his canals.

Beyond his death, the Flagstaff Observatory continued to be staffed by pro-canal theorists, who by this time included Carl Lampland and Earl Slipher, both very proficient photographers. Indeed, Slipher was to continue to promote the canals from the observatory until his death in 1969. His last publication on the matter was a large atlas of the Martian surface, complete with canals, that came out in 1962 shortly before the Mariner 4 photographs and was used by NASA in the planning of their Mars programme. The canals issue, however, had been settled to the satisfaction of most astronomers in 1909, and beyond the First World War few people mentioned it except as a matter of curiosity. The matter was finally settled by the photographs returned by spacecraft in the 1960s, although there are still those around today who believe in their existence.

SO WHAT WERE THE CANALS?

It is now reasonably certain that the canals were an optical illusion produced as a result of poor telescope quality, causing finer details on the Martian surface to join together to appear like a network of lines. Green and Maunder were the first to suggest this shortly after Schiaparelli's mapping of the canals in 1877, and they held to their opinion until they were eventually joined by their colleagues some thirty years later. Many of the objections that they raised in the meantime could have resolved the issue much sooner had people bothered to listen, especially to their objection that no two maps of the canals appeared to be very similar, suggesting that the features were not fixed or solid on the Martian surface. It was only the embarrassment of many at Lowell's claims that led to the balance tipping towards Maunder, Green and, later, Antoniadi.

The last piece of significant research done on the canals was by Carl Sagan and Paul Fox, who attempted to use the Mariner 9 photographs to try to explain what the canals actually were.[199] Mariner 9 mapped the entire surface of Mars at a resolution that was capable of distinguishing objects 1 kilometre or more in size, with a few select areas where it was capable of seeing objects of 100 metres. This coverage and detail was more than enough to establish whether or not there were large areas of vegetation or a canal network in existence on Mars's surface.

Sagan and Fox took the Mariner 9 photographs and compared them with a number of the canal maps made by Schiaparelli, Lowell and Slipher in the hope of resolving the phenomenon. They divided the range of theories about the canals into six distinct categories and addressed in turn the likelihood of each one being the cause of the canals. These provide us with a good summary of the canals issue and are listed below:

1. This is the Lowellian theory as previously outlined, i.e. that the canals are vegetation-surrounded waterways constructed by intelligent beings to avert a disaster. The chief subscriber to this theory was Lowell him-

self, but the Mariner 9 survey showed no signs whatsoever of the canals, oases or vegetation. A previous study by Carl Sagan and D. Wallace[200] also proved that there were no significant landmarks that could be attributed to an alien civilisation, although modern proponents of 'The Face on Mars' would disagree with this (see Chapter 14).

2. The canals are cracks or deep valleys on the surface of Mars and therefore produced by normal geological process. This has been proposed by Alfred Wallace,[239] E. Pickering[166] and C. Tombaugh.[230] Similarly, J. Joly[106] suggested that they may be ridges, and other authors, such as Giovanni Schiaparelli,[205] S. Arrhenius,[10] J. Wasiutynski,[240] G. Katterfel'd[112] and E. Hope,[90] proposed geological features, including volcanic dykes or fault escarpments, as the cause. Sagan and Fox found at least two canals that respectively corresponded to a valley and ridge on Mars. The valley, Valles Marineris, is 100 kilometres in width and would have been visible through Earth-bound telescopes in the nineteenth century. The ridge system, called Ceraunius, was approximately 1,500 kilometres long and also would have been visible from Earth.

3. Edward Maunder,[137] Eugene Antoniadi,[9] Carl Sagan[195] and many others all thought that the canals could simply be a whole series of features on the planet's surface, such as craters, that when viewed from Earth would blur together to give the appearance of canals. Only two or three possible analogies could be found between chains of craters and Lowell's mapped canals. Sagan and Fox made a special analysis of two canals, ironically named Lowell and Schiaparelli, and suggested that seasonal sand movement could cause dark areas within them to cover and uncover, possibly mimicking vegetation growth and decline.

4. Craters were again used by J. Plassman,[168] and by P. Oncley and C. Fulmer,[158] but this time it was speculated that radiating rays of ejected molten rock which spread out from craters could have been mistaken for straight waterways. In reply, Sagan pointed out that on the Moon, and other planets, such crater rays are light in colour, not dark, and that the number of these features on the Mars is so minimal that they are not a satisfactory answer to the canal problem.

5. F. A. Gifford[76] suggested that the canals might be the long ridges of enormous desert sand dunes. Dunes were indeed found on Mars by Mariner 9, but they were too small to be seen from Earth.

6. Finally, Carl Sagan and others[197, 198, 199] thought that the canals might simply be the seasonal covering and uncovering of dark surface features on Mars by windblown sediment. This was Sagan's and Fox's favoured theory, but even they admitted that this could only explain another two or three canals at the most. They particularly quoted the example of the

Cerberus region of Mars, which, although reasonably featureless, did have darker regions where fine dust did not appear to have been deposited. These darker regions, which would be covered and uncovered seasonally, could have been responsible for some of the canals.

Additional theories not discussed by Sagan and Fox include a variation on the Lowellian theme where, in 1959, Donald Lee Cyr rather novelly suggested that the canals might be large mats of vegetation growing out of geological cracks in the crust of Mars. Needless to say, the Mariner mission did not find evidence of this either.

At the end of their research Sagan and Fox could explain only 10 to 20 of the 437 canals seen by Lowell and others on the planet's surface. The Mariner 9 mission had finally laid the matter to rest, determining that there are definitely no canals, or even canal-like features, on the surface of Mars. The matter of what exactly it was that Schiaparelli, Lowell, Slipher and others were seeing on Mars has never fully been resolved, although the 'optical illusion' theory of Maunder and Evans is the most likely.

Since the Sagan and Fox survey of 1975 the Viking mission, with its even higher resolution mapping of the entire planet, has come and gone. This again found no evidence of any canals, or features that could explain them, but by then the canals were no longer worth commenting on and had merely become another curiosity in the history of the study of Mars.

The canals debate tarnished the reputation of planetary science for some time, leading to a general decline in the study of Mars during the first half of the twentieth century. Percival Lowell himself is now seen as an over-enthusiastic and interfering amateur whose fortune and personal charm got him far more attention than he should have received. Needless to say, the word 'canal' has not been mentioned with any seriousness by astronomers for several decades. In an interesting postscript, it might be thought that people would have learnt from the errors of the canal theory, but the same mistakes are being made again over a series of Viking photographs purporting to show extraterrestrial buildings in the Cydonia region of Mars. These are discussed more fully in Chapter 14.

COMMUNICATION BREAKDOWN

Before we leave the issue of Mars and its canals, there is one last offshoot to the story that is of relevance to the 'life on Mars' debate. The widespread interest sparked by the possibility of intelligent life on Mars led to a number of ideas for trying to communicate with the extraterrestrial beings. As these suggestions were made before the advent of television, radio and even conventional flight, most of them relied upon creating giant patterns on the Earth's surface that could be seen from Mars in the same way that we could see its canals. Others thought that a better way would be to reflect sunlight up to the planet in regular Morse code-style bursts so that any Martians would know that its origins were artificial. Here are a few of the suggestions made by various turn-of-the-twentieth-century astronomers, and others, for contacting our Martian brethren.

The first suggestion for building a giant mirror to reflect sunlight up to Mars actually came in 1869 from Charles Cros, a colleague of Flammarion's, nearly eight years before Schiaparelli's canal map was published.[52] Others to suggest this idea included Francis Galton in 1892,[75] who favoured mirrors, a Mr Haweis,[82] who favoured electric light, and the Russian Konstantin Tsiolkovskii, who favoured 18 kilometre-wide rotating mirrors in the desert to make firm contact.[232] William Pickering of Lowell's Flagstaff Observatory made a serious bid for $10 million in 1909 to build a ring of 5,000 mirrors, all 18 metres wide, with which he could signal Mars. Commenting on this theory, Wilfred Griffin suggested using searchlights instead. The debate about mirrors and searchlights carried on in the press for some time but faltered after a number of tongue-in-cheek proposals were made by contributors advocating things such as 'drilling a hole through the Earth'. Other methods of communicating were less expensive and generally involved planting large strips of trees, in a geometrically recognisable shape, across barren countryside or using lengths of black cloth on a light background for the same purpose.

Others claimed to have seen signs of the Martians' having communicated with us. The most famous of these was from the electricity pioneer Nikola Tesla, who claimed that he had used the new wireless system to detect a radio message from Mars in 1901 and that he was designing a system by which he could answer back.[226] A more bizarre claim came from an 1895 edition of the *New York Herald* in which it was claimed that surface features on Mars could be observed to form the Hebrew word for God.

Finally, in 1891, there was a point when Martian life looked so certain that a French widow, Cara Goguet Guzman, offered a prize of 100,000 francs for the first person to communicate with extraterrestrials. However, she excluded Mars as being too easy a target to contact. With all the excitement that existed over Martian communication, the astronomer Edward Barnard wrote a story wherein scientists, upon receiving a message from Mars, signalled back the question 'Why do you send us signals?'. The Martians replied, 'We do not speak to you at all, we are signalling Saturn!'[216]

5. THE MEASUREMENT OF MARS

A DECLINE IN INTEREST IN MARS

Although the perihelic opposition of 1909 removed the notions of canals, oases and dying civilisations from the minds of most scientists, many other aspects of Lowell's vision of Mars remained in place until the start of the modern era of space exploration.

Lowell's death coincided with a fall in public interest in the planet, and there were a number of possible reasons for this. It may have been partly due to the embarrassment and controversy that the canal episode caused in the first place. Equally well, it could also have been due to the rise in interest in what is now known as quantum physics and in Einstein's work on general relativity. More than all this is probably the fact that from the end of the First World War onwards, research on Mars concentrated more on making scientific measurements of the planet rather than mapping features on its surface.

Lowell and many other nineteenth century observers were self-taught astronomers who had been given access to very good telescope facilities. They were, on the whole, not trained scientists, and their deductions about Mars were based on their observations of the planet night after night over a number years and decades. By the time of Lowell's death telescopes were reaching the limit of their usefulness for observing Mars, making it difficult for the Victorian 'gentleman' astronomers to contribute anything new with their line drawings of Martian features. Instead, a new generation of astronomers was emerging who were university-trained scientists. These people were less interested in mapping the geography of Mars and more concerned with establishing facts about the planet's geology, atmosphere and climate. The search for life formed a part of these measurements and, surprising as it may seem, people were still highly influenced by Lowell's vision of Mars as a habitable planet, albeit minus the canals.

Lowell's final vision of Mars was of a planet with an equable temperature, an atmosphere of similar pressure and composition to our own, seasonal meltwater from the polar caps flowing down a network of canals, and dark dry ocean bottoms filled with vegetation. Some of these beliefs were lost, although others were maintained until the arrival of the space age.

THE SCIENTIFIC ANALYSIS BEGINS

Before discussing any further results, it is worth explaining the nature of spectral analysis, a tool used by astronomers in the study of the planets, including Mars. Spectral analysis was, and still is, a commonly used method of determining the chemical composition of astronomical objects. The technique resulted from the discovery that light is not just white, but can be split into a spectrum of colours using a prism to produce a rainbow. In 1814 it was discovered that the light from the stars and planets was not complete but that there were tiny black lines in their spectra. After experimentation, it was found that the 'missing gaps' in the spectrum of a star or planet corresponded to the wavelength of certain elements, and that by measuring these gaps it was possible to work out the chemical composition of the planet or star from which the light originated. In the case of Mars, the light we receive would have travelled through the Martian atmosphere, allowing astronomers to estimate its composition using a specialised instrument called a spectroscope.

This works very well, but there is a big problem. Any light received from Mars would also have to travel through the Earth's atmosphere before it could be measured in an observatory. Thus any spectroscope measurements contained a mixture of Earth's and Mars's atmospheric components, and before the Martian atmosphere could be known the Earth component had to be subtracted. This is not easy, and so, depending on the calculation made, the estimates of Martian atmospheric conditions varied considerably from astronomer to astronomer.

The first attempts at making measurements on Mars were for the detection of water vapour in the decade before Schiaparelli's 1877 map. William Huggens, Jules Janssen and Hermann Vogel, all European astronomers, had used comparisons between the colour spectrum of the Moon, which had no water or atmosphere, and that of Mars. Differences between the two led all three astronomers to declare that water vapour was present in Mars's atmosphere.

It was nearly twenty years later, in 1894, before these findings were disputed by the American astronomer William Campbell. He used the same spectral comparisons between the Moon and Mars, but this time concluded that neither water vapour or oxygen was present in the Martian atmosphere.[34] The presence of water vapour was a crucial part of the canal theory, and in 1903 two of Lowell's assistants, Slipher and Lampland, repeated

Campbell's experiments to try to disprove his findings. They received some results, but unfortunately the two astronomers declared that their equipment was not sensitive enough to make the necessary measurements and the issue was dropped. Slipher tried again in 1908 and this time managed to get the readings he required: he declared that water vapour was to be found on Mars.[9]

Campbell was, by this time, very vocal in his opposition to Lowell's Martian theories and said of Slipher's measurements: 'I have had no confidence in their [Slipher's and Lombard's] evidence of water vapour . . . The critical band lies just in the beginning of the region where Slipher's plates fall off exceedingly rapidly in sensitiveness.'[35]

To prove his point, Campbell used the opportunity of the perihelic opposition of 1909 to climb to the summit of the 4,400-metre high Mount Whitney, taking all his telescopic equipment with him. Here he again took spectral measurements of Mars which, in the clean, dry and clear air above the mountain, would be more precise and accurate than previous measurements. This time he compared the levels of water vapour in the dry air around him with those on Mars and again concluded that Mars had little, if any, water vapour in comparison to Earth. Although by this time the canals were out of favour with astronomers, many of Lowell's other ideas, especially water on Mars, were not and Campbell's results were not widely trusted.

Some years later, in 1934, Walter Adams and Theodore Dunham made further spectral analyses of Mars in search of oxygen and water in its atmosphere. Instead of using conventional spectral analysis, they used a method that was first proposed by Lowell himself, although he never actually tested it. Ironically for Lowellians, their results were entirely negative and it was generally accepted that Mars's atmosphere was lacking in oxygen and water vapour compared to that of Earth.[1, 2] The debate was settled in 1963 when an extensive study of the Martian atmosphere by Audouin Dollfus,[60] which is discussed more fully later, revealed only trace amounts of water vapour in the atmosphere.

Water was, and still is, an acknowledged prerequisite for the existence of life on other planets and was definitely needed on Mars in order that people might explain the presence of both the canals and the wide areas of vegetation. For 'pro-lifers' to admit to Campbell's, Adams's and Dunham's measurements was as good as declaring Mars a dead planet. Fortunately, the problems associated with the accuracy of spectral measurements left room for error, so that it was still possible for at least some water vapour to be present in the atmosphere. There was also the question of the clouds and the polar ice caps. Surely these were signs that water, in some form, was present on Mars? These two topics were also dealt with extensively in the post-Lowellian study of Mars.

The clouds themselves were first observed by Schiaparelli during the favourable opposition of 1877 and were next seen with clarity during the opposition of 1909. During the summer of the latter year Antoniadi observed that for a number of days Mars was obscured by periodic and patchy yellow clouds.[9] He observed the same phenomenon in 1911 and 1924, and explained them not as conventional water vapour clouds but instead as huge clouds of fine dust and sand lifted high into the Martian atmosphere by convection winds heated by the Sun. The same theory was proposed by G. Kuiper, who saw at first hand the so-called Great Martian Dust Storm of 1956 and used it to explain away the theory that vegetation existed on the planet's surface,[121] suggesting instead that the seasonal waves of darkening, interpreted as vegetation growth, were in fact dust storms.

The nature of the polar caps was less problematic as it had always been assumed, from their first observation by Cassini in 1666, that they were composed of frozen water, few people disagreeing with this hypothesis. It was William Herschel, in 1784, who first formally proposed that they were made of frozen water.[85] Between this and the deployment of the first space probes to the planet only four authors, J. Joly, A. Ranyard, George Stoney and Alfred Wallace, proposed anything different. They, working at the turn of the twentieth century, thought that the polar caps could be made of frozen carbon dioxide gas, an idea that was not entertained then or for several decades afterwards and was later proved false by the first successful Mariner mission.

To Lowell the ice caps were an integral part of his Martian theory. After all, it was the seasonal melting of the caps that provided enough water to fill his canals and therefore bring water to his dying equatorial civilisation. However, although few observers disagreed with the frozen water composition of the ice caps, the idea of their seasonal melting was not so readily accepted.

Lowell's seasonal melting was based on his observing a seasonal wave of darkening that spread from the caps toward the equator of the planet. This was, he assumed, a wave of vegetation growing in the path of the migrating polar meltwater. In fairness, the wave of darkening was seen by many astronomers throughout the years, including Antoniadi, who, like Lowell, placed the explanation on surface vegetation. He thought that the vegetation itself may be broadleaf in nature, based on a colour change he noted from dark green in the spring to an autumnal yellow at the end of the summer.[9] In contrast, Kuiper thought that any vegetation was likely to be moss-like in nature to take advantage of the lack of moisture.[122]

VEGETATION

There was surprisingly little opposition to the theory that vegetation might cover the surface of Mars, and there was still a strong belief in this up until the early 1960s. Indeed, scientific confirmation of this fact came from W. Swinton, who, in 1959, used spectral measurements of Mars to confirm that the darker surface areas were indeed composed of organic matter and were therefore liable to be plants.[174] It was thought reasonably certain that some form of plant life, however primitive, would be found on the surface of Mars when the first probes arrived.

Another equally important factor for the survival of vegetation, or any life, on the Martian surface was the need for protection from the fierce interstellar ultraviolet light. In unprotected doses, ultraviolet light is extremely dangerous and destructive to organic matter, causing it to breakdown rapidly. Think of being sunburnt on our relatively ultraviolet-protected planet. On Earth a layer of ozone gas in our upper atmosphere acts as an ultraviolet screen, filtering out the worst of the radiation and allowing life to thrive unsheltered on the planet's surface. Such a system would be needed on Mars if life were to be able to exist there, particularly as there are few clouds to help block out the harmful Sun's rays. For vegetation to be able to survive, the pro-life astronomers needed to find a Martian equivalent to Earth's ozone layer.

Again the 1909 opposition provided the necessary evidence. It was Carl Lampland who in that year used coloured filters at the Lowell observatory to deduce that short-wave ultraviolet light was being filtered out by the Martian atmosphere before its arrival at the planet's surface. The nature of this layer was not known, but nonetheless it was affectionately given the nickname 'blue layer' or 'violet layer' because of the coloured filters that were used to observe its presence. Divergent theories about what exactly the blue layer was varied from its being ice crystals[120] to its being fine dust held aloft in the upper atmosphere.[159]

The blue layer was assumed to be permanent until further work on the issue was done by Earl Slipher at the Lowell observatory in Flagstaff. In measurements taken in 1926 and 1928, Slipher found that the blue layer would sometimes be absent from the planet, and he proposed that it might be seasonal or even occasional in its occurrence.[219] He again measured it in 1937 and was this time quite certain that it varied in intensity and would sometimes be absent for long periods of time—something that was a great problem to a pro-Lowellian such as himself who knew that without it any vegetation, as we know it, would be burnt to a crisp within seconds.

Luckily another astronomer, Seymour Hess, came to his rescue and, rather than accepting a Martian surface bathed in ultraviolet light, proposed that vegetative growth would simply halt during periods when the blue layer was absent and resume again afterwards. He backed this up with observations of the wave of darkening, attributed to vegetative growth, coming to a halt when the blue layer cleared, only to resume again when the blue layer returned. This, if anything, only presented a stronger argument for the existence of vegetation on Mars. We now know that the effect being observed in the blue layer was indeed caused by airborne dust scattering light back from the planet's surface. However, unlike Opik's supposition, this dust does nothing to block ultraviolet light from reaching the planet's surface, making the existence of vegetation there all but impossible. This explanation was not discovered until 1978[228] and so did not affect the belief in either the blue layer or vegetation in the pre-Mariner era. Nonetheless, there were still dissenters to the vegetation theory during this time.

The presence of vegetation on Mars was traditionally deduced by the observation of the wave of darkening that would periodically sweep across the planet. In order to deny the existence of vegetation, the wave of darkening had to explained by some other means. The first alternative explanation came in 1912 when the chemist Svante Arrhenius, a noted opponent of Lowell, suggested that the Martian surface could be made of salts which would darken when the seasonal meltwater ran over them.[10] This was not given any consideration at all at the time, and we have to wait until the 1950s, when massive cracks were already beginning to show in the Lowellian vision of Mars being a hospitable place, before further suggestions were put forward.

American astronomer Dean McLaughlin suggested that the darkening could be due to volcanoes periodically spewing tonnes of black ash into the atmosphere.[140] By this time, other astronomers had speculated about the presence of active volcanoes on Mars and even claimed to have seen them through their telescopes. However, the low atmospheric water vapour content measured by Campbell and others worked against the existence on Mars of active volcanoes, which can pump huge quantities of water into the atmosphere. McLaughlin's suggestion that the volcanoes erupted seasonally was also contrary to observed volcanic activity on Earth and so the theory was discounted.

In 1956, as the volcano theory was still under active debate, Mars came into close opposition again and solved the vegetation mystery for us, although this was not fully acknowledged by many astronomers for some years. That year the planet was engulfed by a huge dust storm which obliterated everything from view for a number of weeks. Based on their observations of this storm, two astronomers independently arrived at a new theory to explain the wave of darkening. These observers, G. Kuiper and

V. Sharanov, mused that if enough dust could be moved into the atmosphere to cover the entire planet then perhaps the wave of darkening was due to the covering and uncovering of darker surface regions of the planet. Carl Sagan and Paul Fox were later to make a similar claim in relation to the canals debate (see Chapter 4), and this is now accepted to be the cause of wave of darkening.

The other vegetative regions outlined by Lowell, and accepted by others, were the dark regions of Mars that he envisaged to be ancient sea beds full of plants clinging on to the remaining water there. This was questioned by Opik, who argued that the dark regions were more likely to be elevated features, not depressions, on the planet's surface because they were never observed to fill with dust after storms.[159] It was argued back that the reason the dark areas did not appear to fill with dust was because the vegetation in them would grow through the dust to reach the light once more. Just before the Mariner 4 mission took off, radar analysis of Mars showed that Opik had been correct and that the dark areas were in fact elevated plateaux that were scoured of surface dust by the Martian winds.

Apart from the water vapour, blue layer, vegetation and polar caps, there were three other important Lowellian theories that came under close scrutiny during this pre-space age era.

ATMOSPHERIC MEASUREMENTS

For reasons stated in the previous chapter, Lowell regarded Mars's temperature as being 'as warm as the south of England', with an atmospheric composition and pressure similar to that of Earth.[130] Both these were important prerequisites to the existence of life on Mars as a thick atmosphere was necessary for the support of liquid water and the presence of oxygen was considered, then, to be necessary for life to be able to survive.

Atmospheric measurements, other than those concerning water vapour, were made in 1927 by D. H. Menzel, W. W. Coblentz and C. O. Lampland,[49] who attached a heat-sensitive instrument to their telescope to measure any heat radiating from the planet's surface. They estimated that the temperature at the equator was between 15 and 30°C whilst at the poles it was –50 to –70°C. These temperatures are not that dissimilar to those on Earth and were certainly warm enough for life to exist in the tropical and temperate regions of Mars. E. Pettit and R. Richardson, using identical methods, obtained the same results in 1954.[163] These estimated temperatures were believed to be roughly accurate until the results of the Mariner 4 probe came back to Earth, and did much to encourage the belief that both life and water could exist comfortably on Mars's surface.

Estimates of the gaseous composition of Mars's atmosphere were much more varied, and although Campbell,[34] and Walter and Dunham,[152] had concluded that oxygen was absent from the atmosphere they were generally ignored, and it was not until after the Second World War that the first proper estimates of atmospheric composition were made.

In 1947 G. P. Kuiper used the powerful McDonald Observatory telescope to perform a full spectral analysis of Mars's atmosphere.[122] By making comparisons with the Moon and with Earth's atmosphere, he estimated that Mars's atmosphere contained 0.06% carbon dioxide. This was approximately twice the amount as on Earth and did not unduly worry the pro-life community, who still felt that there was plenty of room in the atmosphere for oxygen, nitrogen and other gases to exist. Another estimate, by G. de Vaucoulers in 1950, had the atmosphere as nitrogen (98.5%), argon (1.2%), carbon dioxide (0.25%) and oxygen (0.1%), which again did not worry people too much as nitrogen is an essential part of the biological process on Earth, particularly in plants.

Looking retrospectively, the end of Lowellian Mars actually came in 1963 with the publication of a series of three papers, one by Audouin Dollfus and two by H. Spinard, G. Munch and L. Kaplan. All three papers dealt with spectral measurements taken of Mars which were again used to reconstruct the planet's atmospheric composition. Two of these papers dealt specifically with the question of water vapour in the atmosphere. Dollfus detected almost none,[60] whilst Spinard, Munch and Kaplan[221] went further and calculated that if all the water vapour in Mars's atmosphere were condensed out then it would form a layer only 0.0014 of a millimetre deep across the planet's surface. This figure is less than a thousandth of the amount of water available in the Sahara Desert on Earth. Clearly, they concluded, Mars was a very dry planet indeed.

Spinard's, Munch's and Kaplan's second paper dealt with the question of atmospheric pressure.[110] Lowell needed a high atmospheric pressure to allow liquid water to exist on the Martian surface and so that heat could be retained by the atmosphere. His estimate of 87 millibars[131] was based on the reflectivity of light from certain parts of the Martian surface and was backed up by Earl Slipher, who calculated it to be between 83 and 89 millibars.[219] G. de Valcoleurs[236] came to the same figure with his measurement of 85 millibars. In contrast to all these, Spinard et al used infrared radiation to give an average estimated pressure of no more than 25 millibars.[221]

By this time the 'space race' was under way and it was apparent that sooner, rather than later, probes would reach Mars to photograph and measure its characteristics for themselves. Faced with the embarrassing possibility that these probes could find a world totally different from the one envisaged by Lowell and others, scientists began to hedge their bets about the possibility of life on Mars. There were few scientists who did not believe that life would be found on Mars, but in the early 1960s many scaled down their predictions from huge vegetation-covered areas to more modest theories concerning lichens and bacteria. By 1965 Lowellian Mars was no longer being promoted by astronomers or planetary scientists, even though some still believed in its possibility. Although many people anticipated that the first space probes would reveal many unexpected features on Mars, the first results left many people shocked and surprised and many others with egg on their faces.

PART TWO

VISITING
THE RED PLANET

6. A RACE TO MARS

RUSSIAN FAILURES

The space race officially began in October 1957 when the Russians launched Sputnik I, an artificial satellite, into a temporary low orbit around Earth. Ten months later the Americans founded the National Aeronautics Space Administration (NASA), placing the two superpower nations in direct competition with each other in the need to claim political victories in the exploration of space.

The Russians had immediate plans to try to send probes towards our two neighbouring planets, Venus and Mars, and initiated a programme under the leadership of S. P. Korolev and M. V. Keldish. The thoughts of life on Mars, and in particular the presence of vegetation, drove the planning behind their early Mars probes. Instrumentation capable of detecting the presence of organic molecules on Mars's surface was developed at Moscow University and incorporated into a probe that was launched on 14 October 1960, long before the Americans were ready for their first attempt on the planet. Fortunately for NASA, it did not need to worry overly about early Russian attempts to get to the Red Planet for the probe failed to leave the Earth's atmosphere. This heralded a whole string of disasters, on both sides of the Iron Curtain, for probes designed to visit Mars in the early 1960s.

Following this first disaster, the next window of opportunity for a launch at Mars was in the autumn of 1962. The Russians duly launched three further probes at the planet, on 24 October and on 1 and 4 November. Of these, only the 1 November probe managed to leave the Earth's orbit and head on target towards Mars. It was duly given the name Mars 1, its chief aim being to try to capture a number of colour-filtered photographs of Mars as it flew past. Some of these photographs were designed to measure some old Lowellian-based theories, such as the presence of vegetation and the blue layer, as well as more conventional things such as radioactivity and magnetism from the planet. On its way to the planet the probe took continual readings and beamed back a total of sixty-one sets of interstellar measurements to Earth. However, two months before its was due to pass Mars a problem with the orientation of the probe's antennae led to contact being lost, and Mars 1 sailed passed the planet, on June 19 1963, in total silence.

Before describing the entry of the Americans into the race for Mars, it is worth pointing out a major constraint that both superpowers had in getting man-made objects to Mars. Just as the telescopic observers were limited to observing Mars during its two-year opposition cycle (see Chapter 1), so the launching of probes can also only take place approximately every two years. This has to do with the practicality of getting a probe to travel between two planets moving along different orbital paths around the Sun. The most economic method of doing this was devised in the 1920s by the German engineer W. Hohmann, who calculated what is now known as the Hohmann transfer ellipse. This is essentially a calculated path which a probe must follow to travel between two planets efficiently. In the case of Mars, this involves sending the probe out from Earth whilst Mars's orbit is still behind that of Earth's. As the probe travels further out from Earth, Mars will start to catch it up until they eventually meet as Mars's orbit crosses the outward path of the probe. Using the Hohmann transfer ellipse limits the launching of probes to a relatively small window in time every 26 months.

MARINER 4

And so it was that, 26 months after the Russians' previous failed attempts, the Americans finally entered the race for Mars in 1964. In November of that year the Americans launched two probes (on the 5th and the 28th) and the Russians just one. There must have been an eerie sense of unease amongst the NASA scientists as the first of its probes, Mariner 3, failed to separate from its protective plastic shroud and went into a redundant orbit about the Sun. The problem was corrected by the time Mariner 4 lifted off and this probe headed successfully away from the Earth towards Mars. The Russians were again not so lucky, for although their probe, named Zond 2, left the Earth's atmosphere, radio contact was lost with it only five months later. This was their fifth unsuccessful attempt in only four years.

All hopes were now pinned on Mariner 4. This probe was equipped with a variety of experiments which were designed to measure dust, energy and magnetism rather in the inner solar system than specifically on Mars. It did have on board a camera and a magnetometer that were capable of being used on Mars. As the scientific world awaited Mariner 4's flight past Mars, there was considerable debate as to what exactly the probe would beam back from the Red Planet.

The previous chapter outlined the decline in the belief of Lowell's vision of Mars throughout the twentieth century, a belief that really only subsided in the decade preceding the launch of Mariner 4. A comprehensive snapshot of what exactly scientists did expect to find on Mars can be seen in a report commissioned by the Space Science Board of the National Academy of Sciences. The report is the conclusion of a series of meetings, set up by NASA, that took place between the summer of 1964 and October 1965 specially to discuss the possibility of life on Mars. All the major Western exobiologists at the time took part in this series of meetings, and the report itself was assembled just as the Mariner 4 results came back to Earth. It therefore provides the best outline of what was expected of Mars in terms of life and environmental conditions. The following are quotations from the report's pre-Mariner conclusions:

'Mars has retained an atmosphere, although it is thin: present estimates of pressure at the surface range from 10 to 80 millibars. The major constituents are unidentified, but are thought to be nitrogen and argon. Carbon dioxide has been identified spectroscopically and its proportion estimated to lie between 5 per cent and 30 per cent by volume. Water vapour has also been detected spectroscopically as a minor atmospheric constituent.

'The intensity of ultraviolet radiation at the Martian surface may be high by comparison with Earth, but . . . some models of the composition of the atmosphere allow for effective shielding.

'Surface temperatures overlap the range on Earth: they have a daily high of +30°C with a diurnal range of about 100°C.

'Our knowledge of what lies between the polar caps is limited to the distinction between so-called "dark" and "bright" areas. The latter are usually considered deserts [and] the dark areas . . . an optical illusion due to contrast with the orange "bright" areas.'[167]

This expectation of the environmental conditions of Mars is a post-Lowellian one although it still erred heavily in favour of conditions that would be able to support life on the planet's surface. The degree to which life was expected to be found on the planet can be judged by a second extract from the same report's conclusion:

'Biological interest continues to centre on the "dark" areas. In several respects they exhibit the kind of seasonal change one would expect were they due to the presence of organisms absent in the "bright" (desert) areas . . . Infrared absorption features have been attributed to the dark areas, suggesting abundant hydrogen-carbon bonds there, but more recent analysis throws great doubt on this interpretation, leaving us with no definite information, one way or the other, about the existence and distribution of organic matter.

'. . . we find no compelling evidence that Mars could not support life even of a kind similar to our own. Were [limited] oxygen present . . . a fully aerobic respiration would be possible. Some terrestrial organisms have already been shown to survive freeze-thaw cycles of +30 to −70°C . . . extremely low humidities [and a] strong flux of ultraviolet light.

'It is likely that there exist . . . places where the extremes of temperature, aridity and adverse irradiation are markedly ameliorated. Even the presence of water in the liquid phase is perhaps not unlikely, if only transiently, by season, in the subsoil.

'Given all the evidence presently available, we believe it entirely reasonable that Mars is inhabited with living organisms and that life independently originated there.'[167]

Thus, the collective opinion of some of the world expects in Martian studies was that they would expect to find life on Mars, possibly even the large vegetated areas predicted by Lowell, and that the environmental conditions on the surface would be equable enough to support even some forms of terrestrial life. Other astronomers, who did not contribute to the above study, still expected to find forests, seas and even civilisations on the planet's surface.

Apart from the loss of two of its instruments (the plasma probe and the Geiger counter), Mariner 4 was working well as it approached Mars during the summer of 1965. On 14 July it reached its closest fly-by point, which placed it 9,850 kilometres above Mars's surface, and instructions were sent from Earth to operate the spacecraft's television camera. As it swung past the planet its camera took a series of 22 photographs in an arc movement across the sunlit side of Mars from approximately 37° north through the equator towards the south pole, finishing around about 35° south. The photographs were taken on negative film that was processed internally inside Mariner and then placed in the equivalent of a fax machine. Radio transmission of the digitally converted photographs took ten solid days to beam back to Earth, a chronically slow pace in comparison to today's almost instantaneous transmissions of much large amounts of information. NASA held its breath whilst each photograph was painstakingly reconstructed from the hundreds of feet of ticker tape that were received from Mariner 4.

The pictures themselves were, for that time, of reasonable quality, consisting of 200 lines of 200 pixels each and giving the appearance of a grainy, old-fashioned black-and-white television picture. The first pictures that came through were taken a long way from the planet and showed only an outline of Mars, but as the probe got closer details of the surface itself were revealed. Although most people did not expect to see forests of green trees, they equally did not expect to see the stark, barren and cratered surface that the closest photographs actually showed. To all intents and purposes the Mariner photographs of the surface of Mars showed a world that was not that dissimilar to the Moon.

Of all the features seen in the photographs it was the craters that came as the biggest shock. On the 22 photographs, a total of 300 meteorite impact craters were counted, some of them as large as 120 kilometres in diameter. To geologists the existence of wide-spreading cratering on the surface of Mars meant that the planet was a sterile, inactive place and probably hostile to life. The last major phase of meteorite activity in the solar system took place over 3,800 million years ago and peppered everything, including Earth, the Moon and Mars, with impact craters. On Earth all evidence of this cratering has long since been worn away by wind and water erosion or has been pulled back into the crust by tectonic processes. On the Moon, where there is no wind, water or tectonic activity, the craters have remained and are still visible to this day. The Mariner 4 photographs revealed that the same situation existed on Mars, with no evidence of any of the craters having been eroded or subducted into the crust. This meant that no water erosion was present on the surface and that volcanic activity was unlikely to have occurred. In addition to this, there were no signs of any canals, vegetation, rivers, lakes, oceans or other indications that life was, or could have been, existing on the planet's surface somewhere. The phrase most often used to describe the Mariner 4 photographs was 'moon-like'.

Mariner 4 performed one more vital function on Mars before finally disappearing away from the planet. As the probe vanished behind the planet, radio contact was lost for 54 minutes, and the signal transmitted just before and after this disappearance would have travelled through the atmosphere of Mars and could therefore be used to work out the atmospheric pressure there. The result was much lower than previously expected, between 4.1 and 7.0 millibars, and meant that even if there were water on Mars the atmospheric pressure was below its triple point and it therefore could not exist in its liquid form. Liquid water is a requirement for life on Earth, and without it on Mars the chances of life there seemed very remote. Other measurements taken from the planet by the probe indicated that almost no screening of ultraviolet light took place in the atmosphere, that the temperature varied between −133 and +23°C and that there was no magnetic field surrounding the planet, which indicates a lack of geological activity beneath the crust.

Carl Sagan probably summed up the feelings of most exobiologists about the Mariner 4 results when he wrote:

'Some scientists, including some biologists, have been dismayed when confronted with the apparent inelemeney [i.e. hostility] of the present Martian environment. With global mean temperatures of –63ºC, diurnal temperatures in excess of 100ºC at equatorial latitudes, mean surface pressures hovering uncomfortably near the triple point of water, no detectable oxygen, an ultraviolet flux that delivers the mean lethal dose to typical unprotected micro-organisms in seconds, and a surface at one time widely advertised as 'moon-like,' some initial reserve about the habitability of Mars does seem to be in order.'[196]

The Mariner 4 results heralded a new age in the speculation about the possibility of life on Mars. Gone were any notions of complex multi-celled organisms such as plants, lichens or algae covering large areas of the planet, and instead thoughts of life were scaled down to theories about the possibility of simpler organisms, such as bacteria, living in specialised environmental niches. All these theories are still being proposed today. They are discussed more fully in Chapter 9, and, rather than duplicate this debate, we shall here concentrate on the results provided by the Mars missions and their implications for life.

MARINER 6 AND 7

No new spacecraft were launched towards Mars during the available time window in 1967 as both superpowers concentrated their attention on Venus instead. It was to be 1969 before the next wave of craft left Earth for the Red Planet. Almost needless to say, the two unnamed probes sent by the Russians failed even to leave the Earth's orbit after both their booster rockets failed. This pair of disasters led NASA scientist John Casani to conclude that there was a giant galactic ghoul in space deliberately protecting Mars from the multitude of spacecraft that Earth kept throwing at it. In fact, he even suggested that one region of space was responsible in particular by saying: 'Mariner 7 ran into trouble 35 million miles from Earth— a battery got dented, pressure built up and circuits began to arc. A lot of batteries and antennae were wrecked. We began to check the area—no meteoritic activity, no cosmic dust, just clean empty space. Then somebody remembered that this area of space was the same place that the Russian probes, Zond 2 and Mars 1, got fouled up. Further checks showed that two further probes had suffered damage in the Ghoul's lair . . . It could be a coincidence . . .'[40] It was also later suggested that maybe Mars was an interstellar version of the Bermuda Triangle, causing probes to malfunction or lose radio contact as soon as they approached.

Galactic ghoul or not, on 24 February and 27 March the Americans launched two probes, Mariner 6 and 7, successfully. Both probes were considerably heavier and more complex than Mariner 4, their instrument weight being almost tripled. Amongst the instrumentation on board were a high resolution camera, an ultraviolet and infrared spectrometer to measure the composition of the upper and lower atmosphere, an infrared radiometer to measure the surface temperature, an S-band radio occultation experiment to measure atmospheric pressure and a celestial mechanics experiment to measure the gravitational field.

Five months after its launch, on 31 July, Mariner 6 made its approach towards Mars. Its camera was switched on whilst the probe was still 1.2 million kilometres from the planet and a series of 33 photographs was taken and transmitted back to Earth. When it was 560,000 kilometres away another set of photographs was taken and transmitted back. As the spacecraft approached to it closest point to the planet, just 3,431 kilometres away, it took another 25 photographs from 10° west to 265° west along the equatorial region of Mars. These, too, were sent back, and after the various experiments on board had been completed the probe moved away from the planet, its mission having been completed and entirely successful.

Just as NASA scientists were celebrating the success of Mariner 6 its sister probe, which was lagging several days behind, fell silent. Hours of frantic activity followed, during which time it was established that a battery had exploded on board the spacecraft, temporarily disrupting its navigational system. All the Mars probes navigate using the stars as reference points, and fortunately Mariner 7 managed to regain its balance and after a few computer commands to by-pass certain disrupted circuits the probe was functioning as normal once more. Mariner 7, too, took a series of photographs as it approached the planet and 33 pictures were taken of the south polar cap as the probe passed within 3,500 kilometres of the planet on 5 August.

Mariner 4 had previously photographed approximately 1 per cent of Mars's surface at a resolution of 2 kilometres, and by the time Mariner 6 and 7 had finished this total had increased to 10 per cent at the same scale. Fortunately, the quality of the second series of Mariner photographs was much better than that of the 1965 pictures and exobiologists hoped for better signs of life. They were to be disappointed.

The same Moon-like cratered surface was apparent in all the new photographs. There was a brief burst of excitement when, soon after the results had been received on Earth, it was announced that the infrared spectrometer had found traces of methane and ammonia on the outer edges of the southern polar cap. Both of these products are key chemicals in the development and growth of life, and a press release by G. Pimentel at the time stated: 'The qualitative presence of methane and ammonia in the atmosphere gives no direct clue whatsoever concerning its origin. Nevertheless, one cannot restrain the speculation that it may be of biological origin.'[152]

This was reinforced by the calculation that the polar ice caps were made of water ice and not frozen carbon dioxide. Unfortunately, this turned out not to be so, and it was discovered not only that the southern polar cap was made of carbon dioxide ice, but also that it was this ice that had given a false reading on the infrared spectrometer. There had been no methane or ammonia detected on Mars, and so the hunt for life was concentrated elsewhere.

After the disappointment of Mariner 4, the scientific interpretation of the Mariner 6 and 7 results did not once mention the possibility of life. Instead, there was an exhaustive series of papers which put forward climatic, atmospheric and geological models for the planet, none of which seemed very favourable for the existence of surface life. The photographs produced were still of such a low resolution and covered such a small area of Mars that little could be inferred from them except for the largest and starkest of landforms. In terms of the possibility of life on Mars, the exobiology community was at an all-time low, with no canals, vegetation, heat or even water. Instead they had a rather dry and ordinary looking planet covered in dust and meteorite craters.

MARINER 9 VERSUS MARS 2 AND 3

The late spring of 1971 provided the next available window of opportunity for launching probes at Mars, and the two superpowers sent five spacecraft between them. Again, the galactic ghoul ensured that one American probe (Mariner 8) and one Russian (Kosmos 419) never left Earth's orbit. Even so, by late May the Americans had Mariner 9, and the Russians Mars 2 and Mars 3, heading on their way.

There had been considerable advances in technology since the Mariner 6 and 7 visits of 1969, and the intention was to put all three of the probes into an orbit around Mars rather than to have them fly past at a considerable distance. Of the two nations, it was Russia which had the greatest ambitions for its spacecraft, both Mars 2 and Mars 3 having an orbiter part, designed to photograph the Martian surface, and a lander part, which was designed to touch down on the planet's surface. Both landers were equipped with experiments that were to measure the temperature, wind speed, atmospheric pressure and atmospheric composition at the Martian surface. More significantly, both landers also had an experiment on board that would be capable of taking soil samples and analysing them for any organic content to see if there was evidence of past or present life.

The objectives for the Mariner 9 mission had changed considerably after the destruction of Mariner 8. Initially it was intended that Mariner 8 would photographically map 70 per cent of the Martian surface whilst Mariner 9 would concentrate on studying the geology in a narrow band around the equator. In the light of Mariner 8's loss, a compromise was reached and Mariner 9's objective was altered so that it now took over some of Mariner 8's tasks, trying to map as much of the planet's surface as possible. The majority of experiments aboard Mariner 9 were improved versions of those installed in Mariner 6 and 7, including infrared and ultraviolet spectroscopy and S-band occultation. Even so, Mariner 9's features faded into insignificance when compared to the Russian probes travelling with it. It did, however, have one feature which the Russian craft did not, and it was to prove to be of great advantage later on.

On 22 September, two months before any of the probes were due to arrive, telescopes on Earth detected a small dust cloud in the Noachis region of Mars. This cloud grew in size very rapidly and soon became one of the worst dust storms seen on the planet since the great storm of 1956. Almost every feature on the planet's surface was obscured by the thousands of tonnes of dust being whipped through the Martian atmosphere by strong winds. If the probes arrived whilst the storm was continuing, then they would see nothing on the planet's surface at all.

This event had been predicted in February of that year by Charles Capen at the Lowell Observatory, who noted that great dust storms always seemed to occur shortly after the planet's perihelic oppositions; 1971 was just such a year. He wrote that 'a vast atmospheric disturbance could interfere with the first Mariner orbiter spacecraft mission, which is planned to begin reconnaissance of the planet in November'.[36] This was to prove to be very prophetic as the storm continued to rage throughout October and the beginning of November as all three craft approached the planet.

On 10 November, whilst still 800,000 kilometres from Mars, Mariner 9 took a number of test pictures of the planet which revealed almost no detail on the surface save a number of dark spots that would later be identified as volcanoes. The dust storm was still in full swing and the decision was made to shut down the television cameras until the storm passed. Mariner 9 went into orbit around Mars on 14 November to wait patiently for conditions to improve.

Even though Mariner 9's was a less ambitious mission than that of the Russians' probes, the built-in ability to place it on hold gave it, in the circumstances, a huge advantage that was ultimately to save the day. The two Russians probes were entirely automated and unable to have any aspect of their missions altered from Earth, and so it was, on 27 November, that Mars 2 joined the dormant Mariner 9 in orbit around Mars. Unlike Mariner 9, however, this probe immediately started to enact its pre-programmed sequence of events by firstly releasing its lander module for descent to the planet's surface. For an unknown reason, the landing sequence failed to operate properly and the module plummeted through the Martian atmosphere to crash on to the planet's surface. It had, nonetheless, become the first man-made object to reach another planet, and the Russians made much of the fact that the Soviet Union had placed it there. In the meantime, the Mars 2 orbiter took a series of photographs and measurements, all of which were entirely meaningless thanks to the dust storm.

One week later, on 2 December, Mars 3 entered its predetermined orbit and released its lander module. This time the landing sequence worked perfectly with the heat shield being shed, a parachute deployed and, finally, braking thrusters employed. Once on the surface the lander turned on its radio transmitter and transmitted for twenty seconds before falling silent for ever. It is not fully known what went wrong with the lander, but it is suspected that the high wind speeds associated with the storm may have picked up the parachute and dragged the probe sideways through the atmosphere before dropping it on its side on the planet's surface. Again, the Mars 3 orbiter returned results that were next to useless, apart from some temperature readings from the polar regions. All hope was now pinned on Mariner 9.

Finally, over a month after Mariner 9 had entered Mars's orbit, the dust storm cleared and the signal was sent from Earth to mobilise the probe into action. Over the next 349 days it returned 7,329 photographs, covering 80 per cent of the Martian surface, as well as a large number of measurements from the atmosphere and surface conditions. As the photographs were cleaned up and assembled into a large mosaic of the planet, it became apparent that many of the conceptions about Mars, based on the previous Mariner missions, were entirely wrong and that a new era of Martian research was about to begin.

The previous description of Mars as being a featureless and Moon-like planet were blown away as the Mariner 9 pictures revealed a host of new features that were completely missed by the previous missions. Unexpected landforms, such as volcanoes, canyons and, most importantly, dry river beds, were revealed across large sections of the planet. Indeed, the only places that did not seem to have them were those areas that had been previously mapped by Mariners 4, 6 and 7—something that seems to be part of the bad luck associated with the planet in general. In particular, the Mariner 9 photographs revitalised the prospects for life on the planet's surface.

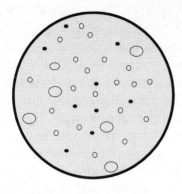
1. Heavy bombardment by meteorites

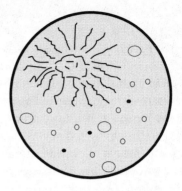
2. Uplift in the Tharsis region

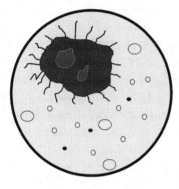
3. Volcanoes cover the surface in lava

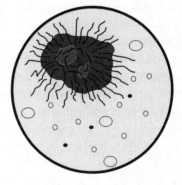
4. Tharsis undergoes uplift again

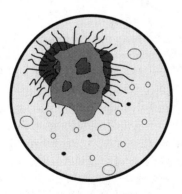
5. Recent volcanoes erupt again

Fig. 5. The geological history of Mars (adapted from Mutch *et al*, 1976).

As has been noted previously, the one overriding concern for exo-biologists was to try to find evidence of liquid water existing or having existed on Mars's surface. The Mariner 9 photographs revealed a vast net-work of channels and tributaries which fan out across the planet's surface in the same manner as rivers do on Earth. Although these rivers showed no sign of having water in them now, they at least proved that liquid water had in the past existed on the planet, suggesting a warmer, wetter and thicker atmosphere than is currently found. The presence of huge volca-noes, including Olympus Mons, the solar system's largest mountain, was also a hopeful sign, showing that Mars was, until recently, a geologically active planet. In addition to the river beds and volcanoes, other insights came to light, including a series of enormous canyons and chaotic flood plains. Moreover, many of the dark, flat regions, which Lowell assumed to be vegetated ocean beds, instead proved to be large plateaux raised high above the planet's surface.

Thus, with the majority of the planet's surface mapped in detail, geolo-gists were able partially to reconstruct the history of the planet's crust. This history began with the heavy bombardment of the Martian surface by meteorites during the earliest existence of the planet. Large areas of these cratered regions still remain on the planet's surface, and it was these that the Mariner 4, 6 and 7 probes photographed to give the moon-like impres-sion of the planet. There was then a massive uplift of the planet's surface in the Tharsis region, located in the temperate areas of the northern hemi-sphere. This uplift resulted in a 4,000-kilometre diameter region of the planet bulging outwards up to 10 kilometres above the height of the rest of the planet. The effect of this bulging was to form large systems of radiating ridges and canyons as the crust cracked under the outward pressure of the bulging crust. Later this was followed by a large number of active volca-noes that covered huge areas of the old cratered surface with highly liquid basaltic lava. Again, further uplift in the Tharsis region caused more crack-ing on these newly formed volcanic plains, to be followed by a more re-cent phase of volcanic activity which produced the massive shield volca-noes such as Olympus Mons which may only be 300 million years old. All this activity has led to there being three distinctive regions on the planet, the ancient southern cratered region, the cracked and fractured northern region and the younger, geologically active area on and surrounding the Tharsis bulge.

A closer study of the dry river channels revealed that they were almost exclusively restricted to the older, heavily cratered terrain in the southern hemisphere of the planet. This, much to people's dismay, indicated that the channels themselves must be at least 3,800–3,900 million years old, although there was evidence that some channels may overlap the younger regions.

Mariner 9 had finally given exobiologists something to be excited about. There had been both surface water and volcanic activity in the past, which meant that life could have evolved during an early, wetter phase and may have adapted to other environments as the water slowly disappeared.

Suddenly the forthcoming Viking mission, whose practicality had been hotly debated, was justified in its objective to search for life. Indeed, a whole host of new theories were proposed as to where life could now exist on Mars. Most of these were concerned with finding niches where life could shelter from the harsh Martian atmosphere. These theories are discussed fully in Chapters 9 and 10 and they served to heighten expectations about the Viking mission.

RUSSIAN MISSIONS AFTER MARINER 9

In actual fact, the Mariner 9 mission was to be the last successful mission to Mars before the Viking probes of 1976. In America the build-up to the Viking mission was prohibitively complex and expensive enough to cause the original proposed launch target in the next available window, in the summer of 1973, to be missed. The Russians, however, used this window to launch another four spacecraft, Mars 4, 5, 6 and 7. All four probes were launched successfully in late July and early August and arrived at Mars in the month between 10 February and 12 March 1974. It was then that things began to go seriously wrong. The first to arrive, Mars 4, suffered braking problems and failed to achieve an orbit around Mars, instead flying past the planet altogether. Mars 5 did achieve its orbit but suffered a power failure after only nine days and 22 orbits around the planet. It did, however, send back sixty useful pictures and a number of measurements in this time. In contrast, Mars 6 and 7 were fly-by missions designed to release lander modules on to the planet's surface. A navigational error caused Mars 7's lander to miss Mars by over 1,300 kilometres, whilst Mars 6's managed to remain active during its atmospheric descent, beaming back information about Mars's atmospheric temperature and pressure before the signal stopped dead 'in direct proximity to the surface',[13] indicating that it probably crashed on to the surface of Mars at some speed.

Out of a total of fifteen missions to Mars in thirteen years, the Russians had only managed one partially successful mission, Mars 5. The 1973 missions had been designed to try and steal the thunder from the planned American Viking mission of 1976, but this last wave of expensive failures seemed to be too much for the Soviets, who decided that they could not compete with the Viking mission and thenceforth concentrated their efforts on the planet Venus with much greater success. It was to be another fifteen years before they looked at Mars again. In the meantime the Americans were busy assembling their Viking probes for what would become one of the most successful and, in terms of extraterrestrial life, controversial planetary missions of all.

7. VIKING

THE VOYAGER YEARS

Almost from the day of its conception, NASA always had ambitions to land an unmanned spacecraft on Mars. Even before the deployment of Mariner 4, NASA had a programme running, called Voyager, whose aim was to try to discover the cost and feasibility of placing a lander module on the Martian surface. It was ultimately abandoned when it was realised that the technology and cost were beyond the America's means in the early 1960s. Nonetheless, NASA firmly held to the idea of developing a Mars lander that could search for life on the planet and convened the Space Science Board of the National Academy of Sciences for a series of meetings to discuss the matter.

These meetings were held in the year preceding the Mariner 4 mission and were to form the basis upon which the Viking mission would be later designed. The committee recognised that the main problem in designing such a lander would be to find a suitable array of experiments that would permit the detection of Martian life, if it existed. A whole range of different experiments were suggested, including ones to detect visual growth or the presence of photosynthesis, respiration, DNA, proteins, enzymes or other metabolic processes. In the end two main conclusions were reached:

'We have reconciled ourselves to the fact that early missions should assume an Earth-like carbon-water type of biochemistry as the most likely basis of any Martian life . . . The fact remains, and dominates any attempt to define landers for detecting life, that no single criterion is fully satisfactory, especially in the interpretation of some negative results. To achieve the previously stated aims of Martian exploration we must employ as mixed a strategy as possible . . . we should not be convinced by negative answers from single [experiments].'[167] In these statements it is possible to see the line of thinking that was ultimately to lead to the construction of the Viking spacecraft.

The unpromising results, in terms of finding life, of Mariners 4, 6 and 7 led to a serious debate amongst NASA staff as to the sense of spending large sums of money sending a probe to search for life on a clearly barren planet. The anti-Viking lobby felt that it would be better to spend the money on a series of smaller missions rather than a single fruitless one. The pro-Viking lobby pointed to the Russians' developments in lander technology and their ambitions on Mars. They also pointed to the fact that, even if life were absent from Mars, the data gathered from the planet's surface would still be invaluable.

The 'pro' lobby won the argument and, despite an air of pessimism about the prospects for life on Mars, the Viking project was granted funds in 1968 and the long process of designing, testing and building the most advanced spacecraft in the history of mankind began. Although it was originally scheduled for launch in 1973, cost and technical problems forced it to be re-scheduled for the summer of 1975. During this time the Russians managed to launch four spacecraft with lander modules at Mars, but, although three of these made it to the planet's surface, the longest any of them operated was 20 seconds (see Chapter 6). As already noted, after this wave of failures the Russians abandoned Mars, paving the way for Viking to reach the planet without any of the superpower rivalry that had marked earlier missions.

So it was that, after the expenditure of $42 million, and having involved a team of 400 people and over eight years' work, on 20 August 1975 Viking 1 sat on top of a Titan 3 rocket at Cape Canaveral, Florida. The launch was flawless, and was followed two weeks later by the equally smooth departure of Viking 2.

The design of the Viking spacecraft itself was a mixture of old and new technology. The orbiter part of the craft was basically an updated version of a Mariner with higher-resolution equipment, including cameras, than had previously been sent to Mars. Beneath the orbiter was attached the lander module, encased in a flat, circular heat shield. When fully assembled, the whole craft resembled a gigantic spinning top.

It was the landers that had taken the bulk of the time and money spent on the project. Although their purpose had been dressed in a number of different ways by NASA, the landers had one central task—to search for signs of life in the Martian topsoil. To this end each contained a number of experiments, which are explained more fully later, designed to test for different signs of life.

After an uneventful journey, Viking 1 arrived in orbit around Mars on 20 June 1976. It remained there for some time, photographing the surface while awaiting the instruction to release its lander module. To keep public interest high in the mission, Viking 1 even opened a science exhibition at the Smithsonian Institute by sending a command to cut a ribbon from Mars.

The Viking 1 lander was scheduled to descend on 4 July, to coincide with the 200th anniversary of the signing of the American Declaration of Independence, but Viking's high-resolution camera showed the original target to be strewn with large boulders. There followed a long period of debate about where to direct the lander. Ideally, NASA wanted to set it down on one of the outwash plains or river channels seen by Mariner 9 as these, having had water in the past, might stand the best chance of yielding life. In the end a plain was chosen to the west of the original site and, on 20 July, the lander separated from the orbiter and began its descent to the surface using a parachute and braking thrusters to slow it down. A hundred things could go wrong, as the Russians well knew, but they did not and cheers went up in Mission Control as the lander transmitted a message confirming that it was alive, well and the right way up. Viking 2 arrived at Mars on 7 August and, after similar problems with its original choice of landing site, its lander touched down safely on the extreme northern site of Utopia, in place of the Cydonia plain as initially intended. Everything had worked perfectly, and panoramic pictures of the red and rock-strewn Martian surface were on every front page across the world. Once the press attention had died down, the true science behind the project could start.

In case of disaster, both landers and orbiters were identical in their design and objectives. The landers had been given a 90-day working life, in which time they were to perform four experiments to test for signs of biological activity. All the experiments received small soil samples placed inside sealed containers by a robotic arm. As some of the experiments needed to share pieces of equipment, such as heaters, they were performed in a set sequence. It was therefore a long time before the full sets of results were gathered in and assessed. Despite more than 40,000 working components in each lander, there were few problems in the execution of any of the landers' functions, although one seismograph did fail to operate. It is, however, the possibility of life that interests us, and the means by which the Viking landers could detect it are now discussed in detail.

THE GAS CHROMATOGRAPH MASS SPECTROMETER

This was the first experiment to be run, and its objective was to detect the presence of any organic compounds within the Martian topsoil and, if they were found, to identify the types of molecule present.

A gas chromatograph mass spectrometer (GCMS) was, in 1976, a relatively new and sophisticated piece of analytical chemistry equipment capable of detecting organic molecules at a level of several parts per billion in the Martian soil. Before its launch to Mars, the GCMS was tested in a variety of situations and performed exceptionally well in all cases. It was understood that some organic contamination from the Earth would exist on the machine when it took off and so, during the long flight to Mars, a test run was carried out on an empty chamber so that the amount of organic material in the chamber would be known before any Martian soil was added. After the initiation of the GCMS test sequence, there was some confusion as to whether any soil had been delivered into the test chamber, but after a photograph of the sample chute revealed sand grains around the edge it was decided to run the test. Unfortunately for the exobiology community, the results from both the Viking 1 and 2 landers detected absolutely no trace of any organic molecules within the Martian topsoil, even allowing for the background measurements.

There has never been any debate about the accuracy of the GCMS results within the scientific community, although, even to the most hardened 'anti-lifers', they were surprising and illogical. Our limited knowledge of planetary soils had led people to expect some types of organic molecules, produced by chemistry rather than biology, to be found on Mars. The negative result was blamed on the high levels of ultraviolet light on Mars, which, it was assumed, must have broken down any carbon-based material on or near the surface.

The team leader of the GCMS project, K. Biemann, offered the only possible explanation as to why any life within the soil might have escaped detection. He proposed that any organisms living on Mars would need to be extremely efficient in order to survive the harsh environmental conditions on the planet's surface. This would probably mean that any organisms would recycle and adsorb any dead material or waste products rather than expel them away from themselves as happens on Earth. Thus, in a low population density of less than one thousand individuals per soil sample, the amount of organic material available for detection would be too low to be measured by the GCMS. This theory is, of course, speculative, and it was not even taken seriously by its proposer. It did, however, offer hope to the pro-life groups and was later used as part of the labelled release result discussion.

Another point that was made was that any organic material within the topsoil of Mars would be very rapidly destroyed by the high levels of ultraviolet light that reach the surface. This would account for the lack of non-biological organic molecules such as the expected formaldehydes, and it led to speculation that even if it were absent from the surface, there may still be life deeper in the soil where there could be more moisture and less ultraviolet radiation (see Chapter 9). The undisputedly negative results of the GCMS results could not totally discount the possibility that life might exist on another planet, although it was be accepted that, in terms of the Viking biological experiments, the experiment was not successful. It was this inability to say a definite yes or no to the results of each experiment that was to be the hallmark of the whole Viking biology programme.

THE GAS EXCHANGE EXPERIMENT (GEX)

The gas exchange experiment was a sealed chamber system designed to analyse the gases given off from a soil sample for any possible biological by-products.

The experiment worked by sealing a soil sample in a canister and then purging the Martian atmosphere out using inert helium gas. After this, nutrients were added to the system in a very fine suspension so that the sample never got fully wet. At periodic intervals gas samples were removed and analysed for their chemical content. Three days later the soil samples were properly soaked in water and the analysis carried out again. Experiments where just nutrients were added were said to be performed in the 'humid mode' and those using water in the 'wet mode'. In addition to variations in temperature, incubation lengths and the addition of water, one sample was deliberately sterilised (i.e. heated to 145°C) before the test was run.

The first humid mode test produced a surprisingly rapid accumulation of oxygen gas after the addition of the water vapour but the levels died down soon afterwards. Could this have been the product of awakened micro-organisms exhaling large quantities of oxygen into the container? Although the signs looked hopeful, the GEx results were to be the start of a long-running argument about whether or not the soil chemistry on Mars could cause life-like reactions to occur in the Viking experiments. This argument was to flare up again and again with each result that came back from the Viking landers.

In the GEx experiment the rapid rise and fall in oxygen levels suggested that a chemical reaction was occurring between the soil and water which quickly burned itself out. In contrast, a biological reaction would be expected to have a much more gentle and sustained rise in gaseous by-products; these would also be expected to exhale more than just pure oxygen. Further analysis showed that the reaction continued until the soil samples had been heated to 145°C, by which time any living organisms should have been destroyed. Running the experiments in the dark also had no effect on the reactions, which meant that, if photosynthesis were responsible, then there was no energy from the Sun with which it could operate. It was therefore fairly certain that a non-biological explanation was needed for the 'humid mode' GEx results.

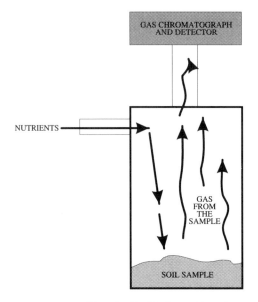

How the Gas Exchange Experiment worked

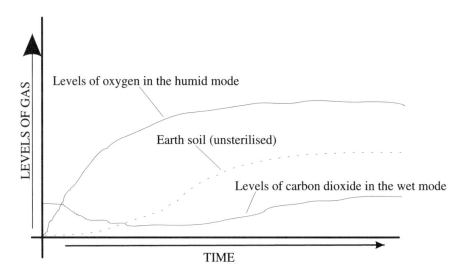

Fig. 6. The nature and results of the Viking Gas Exchange Experiment.

The first suspicion was the possibility that the chemical hydrogen peroxide might be present within the soil. Hydrogen peroxide is a bleach (it is currently used by hairdressers to highlight hair), the presence of which in the cold, dry and irradiated Martian soil had long been suspected. The addition of water vapour to the soil would cause it to break down rapidly into water and oxygen gas, which would produce the steep rise in oxygen levels seen in the GEx. The problem with this is that hydrogen peroxide is chemically unstable when heated and therefore should have broken down when heated to 145°C, stopping the reaction dead. However, this did not occur. It should also have broken down when the samples were incubated for several months, as some samples were, and yet the reaction still occurred after this time. The designer and team leader of the GEx proposed that, instead of hydrogen peroxide, there could be another chemical compound, which he named superoxide, within the soil. This superoxide was given the same qualities as hydrogen peroxide but was considered to be more stable and thus able to withstand excessive heating and lengthy storage.

The second phase of the GEx was performed entirely in the 'wet mode', where the samples had been soaked in water to try to stimulate biological growth. The results from this were to be consistent with the newly proposed superoxide theory. Soon after the nutrient-rich water had been added to the soil sample, about one-third of the carbon dioxide from the sealed chamber immediately disappeared and became dissolved into the soil or the added water. In contrast to the dry mode experiment, no oxygen at all was liberated; in fact, the levels of oxygen in the chamber actually dropped, suggesting that the oxygen too was being adsorbed. The continued incubation of the samples produced a slow and steady release of carbon dioxide until it was back to its starting level, after which time no further reactions were observed. Again, a biological explanation was not sought after the same reaction was seen in a sterilised sample. In addition, the oxygen taken up exactly matched that expected from the amount of ascorbic acid (vitamin C) that was placed in the water as a nutrient. The carbon dioxide uptake was explained by the water causing the proposed superoxides in the soil to take in atmospheric carbon dioxide to form metal oxides or hydroxides. The later slow release of carbon dioxide was explained, rather uncertainly, by iron oxides (i.e. rust) in the soil reacting with dissolved nutrients in the water to liberate the gas. The presence of superoxides in the Martian soil was thus entirely consistent with the non-biological results of the GEx, and from then on the results of the remaining three biological experiments were examined on the assumption that the superoxides existed in the soil samples. However, the results from these other experiments were more problematic than the GEx—and none more so than the next experiment to be run by Viking I after its arrival on the Red Planet.

THE LABELLED RELEASE EXPERIMENT (LR)

Of all the Viking biological experiments, the LR is still the most talked about and the only one to offer a real possibility that microbiological activity was detected within the sampled Martian soil.

The experiment itself was a reasonably simple one wherein a sample of Martian soil was placed in a sealed container and surrounded with a quantity of uncontaminated Martian atmosphere. To this was added a fine mist of nutrient-saturated water vapour designed to encourage any Martian microbes to grow within the sample. The nutrients had been altered to include radioactive atoms of carbon, allowing them to be detected by a Geiger counter later on. The theory was that organisms that took in any of the nutrients and incorporated them into their bodies would also incorporate the radioactive carbon atoms. Later on, via the processes of excretion and respiration, these radioactive atoms should be expelled from the organisms and be detectable by the Geiger counter. An analogy to this would be giving a person an alcoholic drink and then, some hours later, being able to detect the alcohol on the breath using a breathalyser. The alcohol may be gone, but traces of it may remain in the bloodstream and thus be excreted on the breath.

After adding the radioactive nutrients to the soil, the samples were then incubated for up to 52 Martian days whilst a Geiger counter continually measured the amount of radioactive carbon atoms being released from the soil into the atmosphere.

Before its departure for Mars, the LR experiment was run on a variety of terrestrial soils under simulated Martian conditions so that the type of results expected from a biological sample could be known. Bacteria-laden terrestrial soils displayed a steep rise in radioactivity soon after the injection of nutrients, followed by a period of gentler increase before levelling out. By contrast, terrestrial samples that had been heated to 160°C for three hours to sterilise them showed no radioactivity whatsoever on the addition of nutrients, suggesting that they had all been killed.

On Mars, a total of seven test runs were performed by the two landers under varying conditions, including one sample that was sterilised and two that were heated to 50°C.

The LR was the third of the biological experiments to relay its data back to Earth where, in line with the GCMS and GEx, it was expected to produce entirely negative results. However, as most scientists knew, Mars is a very unpredictable planet.

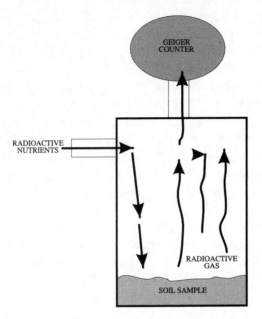

How the Labelled Release Experiment worked

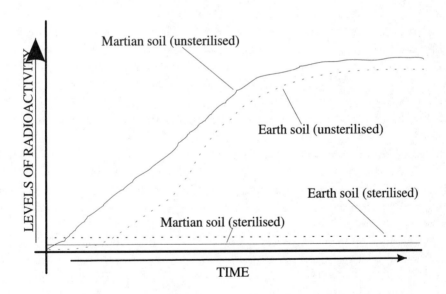

Fig. 7. The nature and results of the Viking
Labelled Release Experiment.

As the scientists watched the results filter in from the first LR test run by Viking 1, they were surprised to see that, instead of a zero reading, the Geiger counter was registering an increasing number of radioactive atoms coming from the soil. After the injection of the first set of nutrients, the level of particles counted climbed steadily for several days before levelling off, indicating a high release of radioactive gas from the soil. However, the addition of a second and third dose of nutrients did not increase the rate of radioactivity detected but actually decreased it slightly, suggesting that some of the radioactive atoms had been re-adsorbed somewhere else in the chamber. When the full readings from the first LR test were viewed, they looked remarkably similar to those resulting from tests carried out on biologically active Earth samples. It all seemed too good to be true.

A second sample was tested but this time was sterilised to 160°C for three hours to kill any organisms. On Earth this had produced no reaction whatsoever. Again, the Martian results mimicked the terrestrial ones, no radioactivity at all being detected from the samples. To be on the safe side, a third test was run under the same conditions as the first (i.e. no sterilisation), producing Earth-like results again. A further test was run after storing a soil sample for four months—which had the same effect as sterilisation—and no radioactivity was detected. To exobiologists the LR looked hopeful and the landing of Viking 2 was awaited with great eagerness. In the meantime, a number of theories, which are discussed later, were put forward to explain the results non-biologically, and accordingly slight changes to the Viking 2 LR experiments were made to help prove or disprove these.

The first LR test on Viking 2 was run under the same conditions as the first Viking 1 test and produced the same result. The next test was carried out after heating the sample to 50°C for three hours. This, on Earth, would severely damage most of the micro-organisms in a soil sample but should leave any chemical reactions unaffected. Thus, if life were responsible for the results, then the levels should be vastly reduced; if soil chemistry were responsible, then the results should remain unaffected.

After this test, the radioactivity did indeed show a marked decline from before, suggesting a biological origin, but a strange fluctuation in the readings made NASA suspect that a malfunction had occurred in the instruments. Another 'normal' test was run to see if there were any differences compared to the original runs; there were not, and it was assumed that the instrumentation was working correctly. Another sample, heated to 50°C, was tested, producing the same reduced levels of radioactivity seen first time around, which again suggested that life might be responsible. With all the results in from the LR test, a furious debate erupted between the pro- and anti-life scientists.

TRYING TO EXPLAIN THE LR

Had the LR experiment been the only one on board the Viking lander craft, then the conclusion would have almost certainly been that life had been detected on Mars. The results obtained were almost identical to the pre-mission criteria for the positive detection of a biologically active carbon cycle involving life on Mars. The problem came when the results were compared with those of the other Viking experiments. Gilbert Levin, the labelled release's designer, was to fight hard against NASA's generally negative attitude towards the results of his experiment, which he considered to have potentially detected life.

One of the main contradictions was between the LR's positive findings and the GCMS's totally negative ones. Scientists asked: How can biological life exist within the Martian soil if no organic material was detected in it? Levin retorted by estimating that on Earth the amount of radioactive carbon dioxide detected in the Mars LR experiments would equate to approximately one million bacteria within the whole sample. Considering that an average temperate Earth soil sample can have hundreds of millions of bacteria per cubic centimetre, this is a very low level indeed. In fact, Levin argued, one million bacteria would be far too low a number for the GCMS to pick up if just the living cells were available for analysis. However, in terrestrial soils the amount of dead organic matter far outweighs the living, and if the same were true on Mars then one million bacteria cells should have been detectable by the amount of waste and dead cells that the community would produce. It was speculated that perhaps Martian microbes might not expel any waste matter from outside their cells, or that any expelled dead matter would be quickly destroyed by the ultraviolet light on Mars, and so the possibility that the LR had detected life could remain despite the GCMS results.

To try to prove this, a series of GCMS and LR tests were set up on Antarctic soil samples.[125] These samples were measured in an identical manner to the Martian ones and the results showed that in some samples the LR showed results identical to those on Mars whilst the GCMS could not detect any organic molecules. A physical examination of the Antarctic samples did indeed find sparse microbiological communities, leading Levin and Stratt to proclaim that 'this finding supports our position that a biological interpretation of the LR experiments must still be considered'. Both these authors felt that they were being bullied into finding a non-biological explanation for the Mars test that they had designed. Having placed doubt on the GCMS readings, they then proceeded to highlight contradictions from the other Viking results and even turned other NASA scientists' arguments against them.

Initial explanations for the LR results again returned to the superoxide theory used to explain the oxygen generation in the GEx experiments. It was suggested that superoxides within the soil were reacting with the radioactive carbon atoms in the nutrients to produce radioactive, and therefore detectable, carbon dioxide gas. However, the same theory had been used to explain the release of oxygen and adsorption of carbon dioxide in the GEx tests. Levin quickly pointed out that superoxides could therefore not be used to explain both the GEx and LR results simultaneously—firstly, because the amount of oxygen released in the GEx experiment was completely incompatible with the quantity of carbon dioxide released in the LR experiment, meaning that two separate chemical reactions were at work; secondly, because the GEx experiment produced the same results after heating the samples to 145°C whilst the LR did not; and thirdly, because the GEx results worked after soil storage for four months while the LR did not. At the end of that particular debate, round one was conceded to Levin.

A further, and novel, explanation was put forward by R. Plumb,[169] who postulated that the gas released was not carbon dioxide at all but carbon monoxide given off by complex chemical reactions between the alkaline, nutrient-rich solution coming into contact with a postulated acidic Martian soil sample. This theory did not stand for long as all the information received from the other experiments suggested that the Martian soil was highly alkaline. Round two to Levin.

Just as it looked as though biology was about to win the day, Levin and Stratt themselves came up with a third and entirely plausible alternative to their own biological interpretation. In addition to the speculation about superoxides, many of the Viking scientists felt certain that the Martian soil contained another oxidising agent, hydrogen peroxide, in significant quantities. Hydrogen peroxide had been ruled out as the cause of other Viking results because of its chemical instability, but this did not mean that it was not present in the soil anyway. It would certainly be capable of producing a rapid burst of carbon dioxide and would have broken down when heated to 50°C and also when stored for four months. It would not, however, be capable of producing the subsequent slow release of carbon dioxide observed over a number of weeks after the initial rapid burst. To explain this, a third oxidising agent (in addition to the superoxides and hydrogen peroxide) was needed which would continue slowly to break down the nutrients into carbon dioxide after the hydrogen peroxide had finished reacting. Unfortunately, nobody could suggest what this third agent might have been. Nonetheless, despite a complete lack of evidence for the speculated third oxide, Levin was initially content with this theory—content, that is, until the results of one of the Viking 2 LR tests came through, where a sample was heated to 50°C.

This test had been designed to test the hydrogen peroxide hypothesis. Heating the soil to 50°C should have been enough to destroy any life within it but would not have been enough to break down the hydrogen peroxide. Thus the results, if caused by hydrogen peroxide, should not have been affected at all, but if they were caused by another means (including life) then the results should have been greatly diminished. As discussed already, the results did indeed show a dramatic drop in radioactivity, shifting the emphasis away from hydrogen peroxide and back towards life again. Levin and Stratt abandoned their previous non-biological theory and erred again towards a biological explanation for their LR results.

In contrast, rather than re-evaluating the evidence, other Viking scientists tried adjusting the existing theory by speculating that under the low atmospheric pressure of Mars the oxides could boil off at a lower temperature, producing the diminished results. On hearing this Levin is reported to have banged his fist on a table and shouted 'You can't keep moving the goalposts!' He later reminded his colleagues that the atmospheric pressure in the sample chamber was artificially high (to prevent the nutrients degrading) and that in any case the amount of peroxides that would be lost through being heated at 50°C for three hours would not account for the drop in radioactivity seen afterwards. After more research he noted that whilst all the other Viking 2 biology results were proportionately lower than their counterparts in Viking 1, his LR results were actually 25 per cent higher. This suggested that a common explanation for all the results, such as various oxides in the soil, was unlikely. Even so, the general conclusion of the Viking team favoured the oxides, which, although speculative, did at least explain most of the results obtained.

Both Levin and Stratt towed the line in the NASA reports on the mission's results, but it is clear from further interviews and published work that the pair still believe that their LR results are best explained biologically.

PYROLYTIC RELEASE (PR)

This final biological experiment was designed around the assumption that any Martian life would be carbon-based, as it is on Earth. On top of this, it was assumed that these organisms would also need to 'fix' carbon from the Martian atmosphere, i.e. breathe in carbon dioxide gas from which the carbon atoms would become incorporated into their bodies in the same

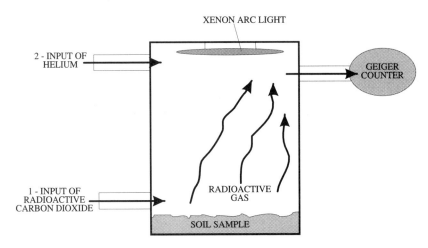

How the Pyrolytic Release Experiment worked

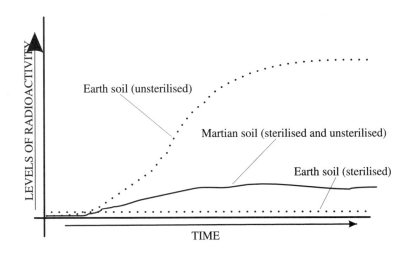

Fig. 8. The nature and results of the Viking Pyrolytic
Release Experiment.

way that oxygen from Earth's atmosphere enters our cells via the blood. The PR experiment was built specifically to look for any carbon molecules that had been breathed in by organisms and had subsequently become part of their body chemistry.

This was done by taking a sealed chamber with a measure of Martian soil in it and introducing a quantity of radioactive carbon dioxide and carbon monoxide gas to try and replicate the Martian atmosphere. Being radioactive, the carbon atoms within the gas could be detected using a Geiger counter and thus distinguished from normal carbon atoms present in the Martian soil or atmosphere. This had the effect of turning each carbon atom into a miniature radio transmitter whose presence could be detected using the Geiger counter. After removing any excess radioactive gas from the chamber, the remaining radioactive carbon atoms could only have been in the soil by having bonded with other molecules to form a solid compound. The Viking scientists hoped that any radioactive carbon atoms measured in the Martian soil would be there through having bonded with molecules inside the body of a living organism.

In detail, the PR experiment took place in a tiny $4cm^3$ chamber containing $0.25cm^3$ of Martian soil. The chamber was sealed and could be artificially lit or darkened to simulate Martian daylight or night conditions. The radioactive carbon dioxide and carbon monoxide (in a ratio differing only slightly from the actual Martian atmosphere) was injected into the chamber in small quantities. It was also optionally possible to inject a small quantity of water vapour into the experiment which could be used to stimulate growth. After injection with the radioactive gases and/or water vapour, the sample was incubated (at between 8° and 26°C) for five Earth days to allow any organisms to breathe in the radioactive gas and retain it inside themselves. The incubation was carried out under artificial day and night conditions.

When the five days were up, any remaining radioactive gas was pumped out of the chamber and the soil analysed for any retained radioactive carbon atoms. The analysis itself was a reasonably complex affair as the carbon atoms would have to be forced out of the resting places and funnelled past a Geiger counter. In practice the only way of doing this was to heat the soil up to a temperature of 635°C, which burnt or vaporised any organic compounds, releasing their carbon atoms back into the atmosphere again. Now liberated from the soil, any gases from the soil were pumped out of the chamber, filtered to remove large molecules, and burnt to produce carbon dioxide gas once more. The carbon dioxide was then pushed past a Geiger counter, where each radioactive carbon atom would be counted individually. Nine successful PR experiments were performed on

Mars (six in Viking 1, three in Viking 2) before a faulty valve in Viking 2 meant that the results in that lander could no longer be relied upon.

After incubation, the results from seven of the nine PR trials showed definite peaks in the quantity of radioactive carbon atoms measured in the soil. Although these peaks were small, they were above those that were measured from sterile soil samples back on Earth and this is suggestive of life in the Martian soil. In fact, it was estimated, using slightly dubious means, that the amount of radioactive carbon detected would be from the equivalent of 1,000 bacteria on Earth.[21] In Earth terms a thousand bacteria in a soil sample the size of that used on Mars would make it practically sterile as a similar portion of soil on Earth could contain millions of bacteria. Norman Horowitz, the experiment's designer and very much in the anti-'life on Mars' camp, said of the results, 'You could have knocked me down with one of those Martian pebbles!'

The varying conditions under which the tests were run found that sterilising the soil, by heating it, before adding any gas or water vapour reduced the measurements to practically zero. This is, again, what one would expect to find if all the organisms in a terrestrial sample had been killed prior to running an experiment. It was also found that measurements were slightly higher under light, as opposed to dark, conditions, possibly suggesting the utilisation of photosynthesis or a similar light-dependent growth factor in the soil. In other tests, the addition of water vapour caused reduced radiation readings, whilst the incubation of samples for 69 and 139 days respectively did not alter the measurements significantly. When compared to the results run on terrestrial soil samples, most of the Viking PR results seemed to suggest that limited amounts of biological activity were occurring in the soils of Mars. However, the initial excitement of this possibility soon gave way to other non-biological theories which were eventually accepted and adopted by the scientists working on the Viking mission.

The problem with a biological origin for the PR measurements was that virtually all the results contradicted aspects of the other biological experiments carried out by the Viking lander. First, the GCMS had detected no trace of organic molecules within the soil, making it difficult to account for the presence of organisms there in the first place (see the discussion earlier). Secondly, the conditions under which the positive measurements occurred were completely different from those seen in the still disputed labelled release experiments. For example, PR measurement still occurred after heating the soil to 90°C, whereas the LR readings died out after 50°C, and the PR experiment was inhibited by the addition of water vapour whereas the LR reaction was increased by it.

After much debate the favoured explanation for the biologically hopeful PR measurements was that a complex reaction was occurring between the carbon dioxide in the atmosphere and chemicals in the Martian soil. This conclusion was easy to reach but harder to prove, as the scientists concerned had no idea what chemicals were present in the Martian soil and no means of testing the soil to find out. They thus had to 'construct' hypothetical chemical reactions which could explain the PR experiments.

The first port of call was, unsurprisingly, the superoxide theory used to explain the gas exchange results so admirably. However, here too there were contradictions when compared to the other Viking results. One of the Viking 1 PR experiments had water added to it and was dried out before the radioactive gases were added and the measurements made. This treatment should have removed all traces of the peroxides before the experiments were run, yet positive results were still obtained. Another theory was needed.

The probable answer came from a study of terrestrial soils undertaken by Jerry Hubbard[118] which revealed that the addition of ultraviolet light to soil samples caused catalytic (i.e. highly reactive) chemicals to react with carbon dioxide to produce organic molecules. A similar situation on Mars would allow the radioactive carbon dioxide gas added to the chamber to combine with chemicals in the soil to form non-biological organic chemicals with radioactive carbon inside them. When later vaporised these would release their radioactive carbon, giving the impression that a biological reaction had occurred. This seemed plausible enough until it was pointed out that all ultraviolet light had been filtered out of the sample chamber and that a number of the experiments had been run in absolute darkness. Clearly, ultraviolet light was not responsible for the observed PR results, and any other options were hampered by our limited knowledge of Martian soil chemistry. Nonetheless, a number of alternatives were proposed, including the chemical assimilation of carbon into polymers, carbonates or other chemicals, but all were contradicted by the observed results in comparison to their test conditions. The conclusion at the end of the Viking mission was that a non-biological explanation was most probable but that a suitable cause could not yet be suggested. In other words, the jury is still out, and, as the head of the Viking biology programme, Harold Klien, wrote, 'at present the explanation of the PR results remains very murky.'[118]

THE LANDER IMAGING EQUIPMENT

Although not commonly mentioned, signs of life were also sought using the onboard camera that both the Viking landers possessed. Although the coverage and resolution of the cameras were limited, experiments on Earth determined that it would be possible to recognise small biological features (down to approximately 2–3 centimetres in size) at distances of up to a few metres away from the lens.

Looking for visual signs of life using the cameras was essentially an exercise in pattern recognition, with the scientists seeking regular-shaped patches on rocks that could represent colonies of organisms and which would change shape over time as the colony grew or shrank. They also looked for other features, such as 'animal' tracks or features with some form of symmetry that may represent an animal. Also, although rather a remote possibility, blurred objects, which may represent something moving in front of the camera, were looked for on the photographs as well. A multitude of pictures was taken by both landers at all times of the day and at all possible angles.

The whole Viking team studied, measured and computer-analysed interesting features on the photographs and concluded that there were no signs of macrobes (large organisms) to be seen.[127] As with many reports from the Viking mission, the door was left open to the possibility of life, the report concluding that 'a model of biology can always be invoked which would have avoided detection by our instruments. For example, Martian photophobes [organisms frightened of light] could always be poised one line scan away, awaiting for the reflected light from the nodding camera mirror to disappear.'

Interestingly, the study of the photographs did reveal a nearby rock exhibiting the letter 'B' which looked as though it had been carved into it. NASA drew no conclusions from this, but, considering the current wave of conspiracy theories about the Moon landings having been faked, it is surprising that this has not been used for similar purposes.

After the negative results of the initial Viking report, Levin again returned to the lander photographs and determined that they showed greenish patches on some of the smaller rocks. He also considered that these patches changed over long periods and that they might represent lichen-like plants on or within the Martian rocks. When lichen-encrusted rocks were photographed on Earth using similar equipment, the results were similar to those seen on Mars.[126] As with 'The Face on Mars' (see Chapter 14), the analysis of remotely taken photographs is subjective and problematic, and the fact that these results have not been highly commented on by others indicates that they may not have been taken too seriously.

THE LIGHT-SCATTERING EXPERIMENT

At the time the missions were planned, there were initially four biological soil analysis experiments for inclusion in the Viking landers. The fourth experiment, called the light-scattering experiment, was devised by NASA biologist Wolf Vishniac, who was keen to design a simple experiment which could only be interpreted in terms of the existence or non-existence of life. The scientist's unusual forename gave rise to the light-scattering experiment becoming known as the Wolf Trap.

The Wolf Trap was a basic test that involved incubating Martian soil in a porous cup that was partially submerged in water. In a chamber beneath the porous cup a light beam would be shone periodically through the liquid and its intensity measured by a detector on the opposite side of the chamber. Any microbe growth within the water would lead to light being blocked from reaching the opposite detector and therefore a lower light intensity would be recorded. The design of the experiment was initially accepted by NASA, who featured it in a 1972 publication outlining the type and nature of the experiments that were to be included in the Viking landers.

Despite its initial approval, the Viking researchers became more dubious about the viability of the Wolf Trap as a means of detecting life on Mars. Arguments were put forward that it was too basic and too presumptive of Martian biology, and in 1973 the project was dropped from the mission.

Vishniac, upset at the rejection of his experiment, later travelled to the Antarctic to test his device on the supposedly sterile soils found in the dry valleys there. It was during the process of collecting samples that Vishniac lost his footing on a ice sheet, fell off a cliff and was killed instantly. Colleagues later tested the Wolf Trap using his Antarctic samples, and it was, ironically, found to be a highly effective means of detecting life in the Antarctic soils. Its effectiveness on Mars may never be known.

THE VIKING ORBITERS

Although the lander part of the Viking mission is the most discussed part of the project in terms of life, it should not be forgotten that, whilst the landers were on the surface, the orbiters were actively taking pictures of the planet's surface.

In the four years that the orbiters remained active, they returned over 60,000 pictures of the Martian surface, the majority at a resolution capable of determining objects of 100–200 metres in size. At this resolution it is obvious that individual or even colonies of organisms were unlikely to be visible on the photographs. There was also no evidence of any large structures attributable to biological actions ('The Face on Mars' excepted; see Chapter 14). The orbiter photographs were nonetheless useful in the 'life on Mars' debate.

In the decades that have passed since the Viking mission, these photographs have been subject to examination by literally hundreds of different scientists specialising in separate areas of planetary research. As a result, we now know a great deal about the past and present processes to have operated on the Martian surface and what they would mean to the prospects of life having evolved or even currently living there.

Whilst individual theories about what and where life may have lived or be living on Mars are discussed in Chapters 9 and 10, but it can be said that the Viking photographs provided detailed evidence about the past movement and location of water on Mars and has outlined areas where frozen water or even hot springs may currently be found. All of these are highly relevant to the 'life on Mars' debate, and the even higher-resolution photograph from the Mars global surveyor will no doubt enhance or detract from the theories about life based on the Viking orbiter photographs.

8. BEYOND VIKING

DID VIKING FIND LIFE?

At the end of the Viking mission, the experiments that had been designed to settle the question of life on Mars had in fact left the issue unsatisfactorily resolved. Of the five direct means of detecting life, three were negative (GCMS, GEx, Imager), one had results that could be either positive or negative (PR) and one had, at face value, results indicative of life (LR). How could such confusion have arisen from a project designed to produce a definite answer about the existence of life on another planet? There are a number of reasons.

Firstly, there is the beauty of viewing the Viking results with the aid of hindsight. It must be borne in mind that when the Viking lander experiments were designed very little was known about the surface of Mars. It was really only when the photographs came back from the Viking orbiters themselves that features of any significance could be discerned on the Martian surface, and by then it was too late to make any alterations to the experiments. As a result, the experiments had to be designed to be simple, broad and able to share vital pieces of equipment such as heaters and injector valves. The technology used in the experiments, although advanced, had to be able to work remotely and off a limited power supply. Given these circumstances, the experiments were actually a miracle of space technology for their time.

Having said this, it was later argued that some of the underlying principles behind the experiment designs were perhaps flawed. For example, the experiments themselves were all designed to find the by-products of life and not actually look for organisms themselves. Bearing in mind our lack of knowledge of any conditions, biological or chemical, in the Martian soil, it is thus not surprising that some of the experiment results got confused with theories about soil chemistry. In this respect it may have been better to devote some of the experiments to a detailed analysis of the soil chemistry so that it could be known whether the proposed superoxides, hydrogen peroxide, metaloperoxides, etc, actually exist. This is again said with the aid of hindsight, but it should nonetheless have been anticipated that complex Martian soil chemistry might mimic many of the simple aims of the biological tests. Other than the imaging equipment, there were also no tests designed to 'look' for signs of life in the soil by using colour saturation analysis or similar techniques. Admittedly, most of these features were prohibitively expensive or too complex to include in a lightweight, remotely operated lander vehicle. Nonetheless, when the next analytical missions depart for Mars they will at least have new objectives, such as the superoxides, to look for.

A second major problem affecting the biological experiments was in NASA's approach to Martian biology. When I was a small boy I was taken around Jersey Zoo by my parents, who took great delight in showing me the stuffed body of the zoo's mascot, the dodo, a large flightless bird that once lived on the island of Mauritius. The dodo became extinct because of mankind's greed, and whilst in the Zoo I was told by my father that the last living dodo had been exported to London Zoo, where, not knowing what it was that dodos ate, the keepers fed it a diet of stones. Not surprisingly, the bird died and the species became extinct. I do not know how true this story is, but I feel that the design of the Viking biological experiments did a similar thing to any organisms on Mars.

All three of the soil analysis experiments were designed to nurture the growth of Martian organisms, but without knowing a thing about what they may or may not require to do this. Instead of designing conservative means of encouraging growth in the soil by providing them with environmental conditions similar to those found in the Martian atmosphere, the biological experiments were all based around creating conditions to which terrestrial organisms would respond well. All the test runs were carried out at higher atmospheric pressures, humidities and temperatures than would be found on Mars, and nutrients were added without the slightest knowledge of what effect they may have had on any Martian organisms. In other words, in the same way that London Zoo assumed that dodos ate stones, so NASA assumed that Martian organisms needed heat, water and nutrients to live and grow.

Norman Horowitz purposely did not use water in one part of his PR experiment as he felt that this would cause any Martian organisms, which must have evolved to survive in an arid environment, to burst open when submerged. In this respect his was the experiment that most closely mimicked Martian environmental conditions. Even so, he still flooded his chamber with an non-Martian ratio of carbon dioxide and carbon monoxide and at a higher pressure and temperature than found on Mars. Similarly, both the LR and GEx experiments involved the use of water, nutrients and artificially high temperatures and pressures. We cannot, therefore, be certain that any life within the soil was not killed from the moment an experiment began.

This speculation about what could or could not kill Martian bacteria raises a third point. All the experiments were designed to detect signs of life as defined by Earthbound biological activity. With the GCMS this was fair enough as carbon, in the form of organic molecules, is a likely contender as a building block of life, few suitable alternatives being known. However, assuming that Martian organisms required water, gaseous exchanges and nutrients in the same manner as terrestrial organisms was very presumptuous. Even if they did require these things, looking for signs of terrestrial biological activity—respiration, the metabolism of carbon, etc—was perhaps, again, too presumptuous of Martian biology.

In fairness to NASA, this is a harsh criticism of a project that was designed with limited technological and financial resources in the early 1970s. As with everything else about Mars, we can only learn by building on the results of previous workers. The Viking lander experiments did not conclusively prove that Mars is a sterile planet and may even have offered the hope of life through the LR results.

Even with today's technology it would be difficult to design another lander that could conclusively prove or disprove the existence of alien life forms about which we know absolutely nothing. The argument can only be settled when samples have been returned to Earth for examination. No such sample return missions are planned until at least 2003—although, in the case of the SNC meteorites, Nature beat us to it (see Chapter 11).

THE END OF VIKING

Once the main biological experiment had finished, both landers continued to function on the Martian surface, using solar power, continually sending back atmospheric readings and photographs. Although they were initially intended to work for ninety days, the Viking 1 lander continued to work for over six Earth years before falling silent on 13 November 1982, whilst Viking 2 fell silent on 11 April 1980. Their orbiters stopped transmitting on 7 August 1980 and 25 July 1978 respectively.

Viking is perhaps the most successful of all the Martian missions to date, and, following the economically turbulent years of the late 1970s and 1980s, was the last big mission to go to the planet until the early 1990s. The generally negative findings of the landers took much of the excitement out of the exobiology debate, but new theories arose which involved life existing, not on the planet's surface, but instead deeper within the soil, where moisture and heat might be more conducive to survival (see Chapter 9).

PHOBOS 1 AND 2 WITH THE MARS OBSERVER

Although the success of the Viking mission led to more advanced plans being forwarded by NASA for funding, the oil crisis and economic recession of the late 1970s and early 1980s took their toll on the space programme budgets. Instead, money was diverted away from Martian missions to the more spectacular interplanetary missions such as Pioneer and Voyager. There was also a feeling that, as with the Moon landings, NASA had beaten the Russians with the success of Viking and that, as the Soviets had lost interest in Mars, the need to score political victories on Mars was lessened. Perhaps the high quality of the orbiter's photographs and the negative biological findings of Viking also reduced the urgency to send further missions. Had definite signs of life been found, then it is reasonably certain that a sample return mission would have been planned soon afterwards.

Indeed, it was to be the Soviets who, in 1988, were next to launch spacecraft at Mars, with a mission designed specifically to photograph and measure Mars's two moons, Phobos and Demios. This mission does not concern us much as its prospects for finding life on Mars were zero. Unfortunately the galactic ghoul struck once more, and, although both probes, named Phobos 1 and 2, made it to Mars, Phobos 1 lost telemetry almost immediately and Phobos 2 followed only three months later. In that time it managed to take a series of useful photographs and measurements, and so the mission was at least partially successful.

In September 1992 the Americans launched a spacecraft called the Mars Observer towards Mars, the first to the planet for nearly eighteen years. The mission consisted purely of an orbiter vehicle that would take very high-resolution photographs of the planet's surface so that research could move beyond that of the Viking orbiters' images. There would have been many implications for the 'life on Mars' issue from this mission as it was hoped that the role of water in Mars's past and present history could be better defined and that signs of subsurface water might be detected. Furthermore, the cult that had grown around human-looking structures on the Cydonia Plain would be able to be resolved (see Chapter 14). However, none of this was to pass as, three days before it was due to enter orbit around Mars, contact was lost with the Mars Observer and was never recovered.

1996 AND THE NEW WAVE OF MARS MISSIONS

Between 1996 and 2005 there are at least eleven missions planned for Mars: four of these are orbiter craft, five are landers and two are yet to be decided. Thus, after the barren period of the 1980s, Mars is firmly back on the scientific agenda once more.

Many new lessons had been learnt about planetary exploration since the 1970s, and, as a result of budget constraints and the end of Cold War rivalry, mission outlines and objectives for Mars had changed radically. Now, instead of sending one expensive probe with many different experiments on board, several smaller missions would be sent over a number of years. This was to avoid the calamities that befell many Russian and American probes, where the loss of a large spacecraft full of instruments would set back exploration by years. Instead, the loss of a smaller probe with only a few instruments on board would be a less serious setback to science, and a replacement probe could be assembled for less money. Furthermore, the objectives for the Mars probes had changed radically from those of the 1970s. The early Mariner craft had been built to find out basic information about Mars, and the Viking mission was meant to refine that knowledge and search for life. All the missions to go to the planet from 1996 onwards are part of an integrated plan to find out whether it will be possible to get a human being on Mars or, at the very least, get some rock samples back to Earth from the planet. Each mission, starting with Pathfinder, is meant to answer questions about the geology, atmosphere and mineral reserves on Mars to see if they can be used as a fuel resource for a future manned or sample return mission. So it was that, in a scenario reminiscent of the Cold War days, in 1996 the Russians prepared to launch their probe, Mars 8 (or Mars '96, as it is sometimes called), and NASA their two probes, Pathfinder and Global Surveyor.

The Russians had spent many years and much money designing and building the Mars 8 spacecraft. The vehicle itself was probably one of the most ambitious ever to be launched at a planet, consisting of an orbiter craft plus two lander vehicles and two further wedge-shaped probes designed to be dropped from the orbiter so that they could embed themselves into the Martian soil to make measurements. Aside from the large number of other scientific instruments on board Mars 8, one of the most significant aspects of the mission was a device designed by the Viking scientist Gilbert Levin. As has been detailed in Chapter 7, it was only Levin's labelled release experiment that showed any vaguely positive biological results from the Viking mission, and ever since he had been badgering NASA to send another lander to Mars to help confirm or deny his earlier findings. NASA had ignored him, but the Russians were happy to include him in their team and on board Mars 8 was a clever experiment designed to test for the presence of any peroxide chemicals in the soil—something that had been used as an explanation for his positive labelled release experiment. This, and the twenty other instruments on board, made Mars 8 an extremely important mission for Martian studies, including the search for life. However, following an all too familiar pattern, Mars 8 joined the long list of failed Mars probes when, only a few hours after its launch in November 1996, the spacecraft could not achieve a high enough orbit and crashed back to Earth. NASA had more luck, and its craft left on their way to Mars on 7 November and 4 December.

PATHFINDER AND ITS SOJOURNER

Although launched second, Pathfinder was the first of the two NASA craft to reach Mars, on 4 July (Independence Day) 1997. The mission itself was a departure from the previous design of NASA craft and consisted solely of a lander with an onboard, remotely operated vehicle called Sojourner. It had also been designed to be as light and as cheap as possible and as a result had no braking thrusters, to save on weight and fuel. Instead the craft braked by using a parachute and then, when only metres from the planet's surface, deploying a series of huge balloons around the craft, cushioning its landing and causing it bounce several times before coming to rest.

By 4 July the world's press had got sufficient publicity for the mission to cause many television stations to cover the landing event live. People held their breath as they waited for confirmation that the probe had survived its unconventional landing and could begin transmitting data. A cheer went round Mission Control as Pathfinder sent back a message confirming that it had landed successfully and was the right way up on the Martian surface.

Pathfinder's objective was essentially a geological one. The main craft was pyramid-shaped and, upon landing, unfurled its sides to form a flat, petal-shaped vehicle on the Martian desert. Contained within this pyramid was the main part of the mission, Sojourner, which was built to roam the Martian surface measuring the chemical composition of any rocks in its path. Sojourner was only 61 centimetres long and 30 centimetres high and was controlled using radio transmissions from Earth, although it did have a limited 'brain' to enable it to avoid any unseen objects such as potholes or boulders. When it found a suitable rock to measure, Sojourner would go forward and press a small x-ray spectrometer against it which would give a detailed chemical breakdown of the rock, allowing geologists to identify it. Although Sojourner's in-built power supply succumbed to the Martian climate after two months, it was able to continue on solar power for some weeks afterwards.

On 5 July the Pathfinder craft sent back the first images from the Mars surface for nearly two decades. In content they were little different from those of the Viking landers, showing a red expanse of sand with a few black boulders strewn across it. Unlike the Viking mission, Pathfinder had been targeted at a very rocky area of Mars in the bottom of an old flood plain on the Ares Vallis, so that a variety of different rock types, swept down from the highlands by the floods, would be available for Sojourner to measure. After an initial problem in trying to get the Sojourner off the main body of the Pathfinder lander, it finally got underway and started to send back geological information from Mars. At the time of writing, the Pathfinder results have only been available for a short period of time, but they have nonetheless already revealed some key pieces of information with regard to the 'life on Mars' debate.

The first thing Sojourner appears to have confirmed is that the canyons, channels and dry river beds seen on Mars's surface were indeed caused by water and not by ice, lava or any other suggested means. It has confirmed this by observing large, 4 metre-high ripples, spaced 20 metres apart, in the terrain near the Pathfinder lander.[114] The shape and size of these ripples, together with the boulders that have been stacked along their edge, match those seen on Earth after sudden, catastrophic floods such as dam bursts and meltwater outpourings. The Ares Vallis was picked as a landing site because it was felt that the valley represented a flood plain, and Pathfinder looks to have proved it right.

The other major finding of the mission has to do with the analysis of rocks performed by the Sojourner vehicle. In its analysis of the closest rocks around the main lander, given such names as 'Barnacle Bill', 'Yogi' and 'Casper', the Sojourner discovered that the rocks had a surprisingly high amount of silicate minerals in their structure.[16] In the case of Barnacle Bill, they accounted for 58 per cent of all the minerals there, which suggested that the rock might be an andesite. This rock type is closely associated with the movement of tectonic plates on Earth and normally forms deep in the crust where substantial amounts of water come into contact with volcanic magma chambers. NASA had predicted that the commonest rock type on Mars would probably be the low-silicate rock basalt, produced when primitive volcanoes, like those on Hawaii, erupt.

The presence of andesite indicates that there must be a considerable quantity of water locked inside the Martian crust somewhere. This water could either be in the porous top 10 kilometres or possibly within the molten mantle beneath. If it is the former, then this brings into question the possibility that, if there is water trapped in the top layers of the crust, there could be microscopic life as well. This possibility is discussed more fully in the next chapter, but what the basic results from the Pathfinder show is that water has played a more important role in the geological history of Mars than has so far been suspected and, as has been pointed out many times in this book, in order to find life as we know it one has first to find water. As NASA researcher David des Marais said, 'For years we've seen the evidence of floods and liquid water on the surface of Mars, but we haven't heard the geochemistry of the story.' He also points out that rocks higher in silica may be better for preserving microfossils on Mars.[48]

The Pathfinder mission, initially expected to last for four weeks, actually continued for over six months, and was declared finished on 12 March 1998 when radio contact could not be regained. Less conventional results inferred from the Pathfinder mission are discussed in Chapter 14.

THE GLOBAL SURVEYOR

The Global Surveyor mission was a partial replacement for the Mars Observer lost in August 1993. Despite being launched nearly a month ahead of Pathfinder, it did not arrive until nearly three months later, finally going into a polar orbit around the planet on 12 September 1997. The craft has a two-year objective whereby it will create planetary maps of Mars's topography and its distribution of minerals and monitor its global weather systems, again all in preparation for a possible manned mission. Due to the length of time taken for the satellite to position itself around Mars, the first pictures were not taken until January 1998, but they immediately confirmed that water had once flowed on the planet. One picture showed unmistakably a dry river bed at the bottom of a 4 kilometre wide canyon, although, as NASA has pointed out, there is still no sign of either tributaries or ancient shorelines. The Surveyor has, at the time of writing, still over a year to go before it reaches its closest point to Mars. Judging from the quality of the few pictures taken so far, the planet will be revealing many more of its secrets and could even show us some likely places to look for evidence of past or present life.

THE FUTURE

There are currently another eight missions planned for Mars, to be launched at two-year intervals. The objectives of many of these missions have yet to be defined, but what is known about them so far is listed below:

1. Missions launched in 1998:
   Planet B (Japan). To measure the upper atmosphere of Mars.
   Mars Surveyor Part I (NASA). An orbiter built to replace the Mars Observer.
2. Missions to be launched in 1999:
   Mars Surveyor Part II (NASA). A lander designed to study the polar regions. There may also be two penetrators similar to those on Mars 8.

Deep Space II (NASA). A lander designed to analyse Martian subsurface soil.

3. Missions to be launched in 2001:
   Mars 2001 (Russia). Undefined. Probably a lander with a vehicle.
   Mars Surveyor 2001 (NASA). Undefined. Probably a lander.

4. Missions to be launched in 2003:
   Mars Surveyor 2003 (NASA). Undefined. Possibly a sample return mission.
   Mars Express (European Space Agency). Undefined and currently unfunded. Probably an orbiter and lander (see below).

5. Missions to be launched in 2005:
   Mars Surveyor 2005 (NASA). Undefined. Probably a sample return mission.

6. Mission to be launched in 2009:
   A possible manned mission is being talked about by NASA.

This list may grow as further nations and agencies become interested in the study of Mars. The European Space Agency are talking about including instruments for detecting signs of life, such as hydrocarbons in the soil, on the lander portion of their proposed Mars Express mission.[136]

Although the direct search for life is a very low priority on the various science agencies' agendas, every piece of firm evidence that is found out about Mars, in particular about the distribution of water, will help us rule out certain possibilities for life whilst perhaps suggesting other places to look for it. When the sample return missions come back to Earth, we should know, one way or the other, whether there is currently life on Mars—or at least in its topsoil.

9. IS THERE LIFE ON MARS?

THE CONDITIONS FOR LIFE

The analysis of the Viking and Pathfinder results brings us up to date on the historical search for life on Mars. It is therefore time to ask what, after centuries of debate and research on the subject, the chances are of finding life on modern day Mars.

This is a question that can really only be answered in relation to what we currently know about Mars's atmosphere, environment, geology and geography. Of all the factors limiting the occurrence of life, it is the environmental conditions on the planet's surface that are most important, and thanks to all the spacecraft results we know just how harsh these are.

In list form, the surface of Mars has no water; an atmospheric pressure averaging 8 millibars, the atmosphere being made of 95% carbon dioxide, 2.7% nitrogen and 0.13% oxygen; a surface temperature varying between –133°C and 23°C; ultraviolet light reaching the surface of the planet almost unscreened; and a soil containing probable superoxidising agents and no organic material, and subject to wind scouring and dust storms.

We have known about these conditions since the 1960s, and it is from this time that the first experiments were conducted to see what types of organism from Earth could survive on Mars. At the time it was assumed, rather naively, that if an Earth organism could be found that would live on Mars, then that is what Martian life would be like.

Most of the organisms tested were bacteria, blue-green algae or lichens, as these were felt best able to survive on Mars. Most of these tests were carried out prior to the detailed climatic and atmospheric measurements taken by Mariner 9 and the Viking probes, so the conditions to which these organisms were subjected were not an accurate representation of the Martian surface. In fact, most of the samples were simply placed in a freezing cold chamber flooded with carbon dioxide gas, although some were also dried out and exposed to ultraviolet light.

The results from these experiments tell us nothing at all about life on Mars, merely that virtually no terrestrial organisms could survive there. It was found that anaerobic blue-green algae[32, 133] and lichens[111, 172, 217] could survive for a short time under Martian conditions, whilst everything else could not. The strangest test I could find was one where, for a 1950s edition of the BBC television programme *The Sky at Night*, Dr Francis Jackson exposed a cactus to a simulated Martian environment for twenty-four hours. At the end of it the cactus looked more like a pancake than a plant, leading Dr Jackson to comment that it had 'a distinctly morning-after look to it.'

At the end of these tests all that was proved was that no known organism could survive prolonged periods of exposure to extreme cold, ultra-violet light or a lack of water and oxygen. Once the true scale of the Martian environment became known and the Viking lander results were accepted, the testing of terrestrial organisms was stopped and new theories were sought as to where life on Mars could survive.

A new quest was initiated by exobiologists to find environments on Earth that would be comparable to the conditions on modern Mars and to see how life there had adapted to them. Thus, by seeing how Earth life adapted to survive in Martian-like conditions, it would be feasible to determine possible survival mechanisms for Martian organisms. As outlined already, the environment on Mars is, compared to that on our own planet, extreme, and therefore evidence for life was sought in the most inhospitable places on Earth.

ANTARCTIC DRY VALLEYS

Even before the Viking landings, it was known that the surface of Mars had no free water, was freezing cold and was whipped by howling winds and dust storms. Life living on or near the surface under these conditions would need to be either extremely tough or else protected from the environment—or, most likely, both. On Earth only one environment comes anywhere near that of the surface of Mars, and this is on the Antarctic continent near the South Pole.

In the early 1960s a number of so called dry valleys were discovered in the Antarctic and were explored biologically. Dry valleys are an extreme environment where the temperature can vary between –60 and 5°C seasonally, where precipitation is extremely low and no free water is available. They are bathed in strong sunlight during the summer and have dry, desiccating winds blowing through them all year round. Between 10 and 15 per cent of all the soil samples taken from dry valley floors are sterile—something quite remarkable in a biologically superabundant world such as Earth. The analogies to the temperature, winds and lack of water on Mars were obvious to most exobiologists, and from 1966 onwards dry valley samples were used to test some of the theories concerning the search for life on Mars.[33] Indeed, in the build-up to the Viking mission, Wolf Vishniac tested his controversial Wolf Trap experiment there and Gilbert Levin used Antarctic samples to help reinforce his claim for a biological experiment for the labelled release results (see Chapter 7).

Biological exploration of the valleys had found evidence of limited amounts of bacteria in the soil, but it was uncertain whether these lived there all year round or were just blown in for the warmer summer months. In 1976, only a short time before the Viking spacecraft arrived at Mars, two researchers made a significant discovery in the Antarctic dry valleys. In contrast to simple bacteria, E. Friedmann and R. Ocampo[74] found an entire ecosystem of primary producing organisms living in the valleys. Making up this ecosystem are photosynthetic bacteria (cyanobacteria), fungi, algae and lichens which actually live within rocks on the valley floor. Such organisms are known as endolithic (meaning rock dwellers), and over the ensuing years the organisation and workings of this miniature ecosystem were discovered and compared to the conditions on Mars.

The endolithic communities were only found in sandstone, living in the pore spaces below the surface of the rock. Here light still penetrates far enough for photosynthesis to take place, but the organisms remain protected by the rock from the desiccating wind, able to exist thanks to the equable environment. Measurements within the pore spaces show that the temperature there can be up to 11°C warmer than outside, allowing small amounts of frost to thaw into liquid water. Other nutrients are derived from minerals and precipitated chemicals, such as iron oxide (rust) and calcium carbonate (limestone), within the rocks and from any nearby dead vegetation. In such a tight and enclosed environment nothing is wasted and only a very small amount of net fixation occurs. Survival is also aided by an absence of grazing animals which would otherwise remove vital material from the ecosystem. Thus the whole ecosystem seems to be stable, despite an apparent lack of moisture, low temperature, etc, making comparisons to possibilities on Mars inviting.

Several authors[227] have used the dry valley endolithic organisms as a basis for their ideas about what life on Mars could be like now. One of the crucial requirements for these organisms is a porous rock within which to live. Their chosen rock type in the Antarctic, sandstone, is a sedimentary rock whose existence has not yet been proved on Mars. However, the presence of fossil lake and river beds suggests that these rocks probably will be found on the planet; even if they are not, some forms of lighter lava, such as pumice stone, would make a good substitute. It has also been suggested that thin cracks and fractures in Martian rocks could be capable of harbouring life.

Endolithic life would stand a better chance of surviving on Mars for a number of reasons. Firstly, such life would be removed from the desiccating Martian wind that is capable of blowing at speeds of 54kph.[86] Dust blown around in this wind could lodge in the cracks or pore spaces, providing valuable nutrients for any life there. A number of authors have suggested alternatives to the carbon and nitrogen cycles seen on Earth, with sulphur and phosphorus (elements known to occur on Mars) used instead.[227] Shielding within the rock could also allow liquid water to exist, even if only seasonally, with the source possibly being the minute layers of frost known occasionally to occur on the outside of rocks during Martian winters. Other organisms may be able to take water directly from the small amounts of vapour in the Martian atmosphere.

There are, however, other problems on Mars besides the need for shelter from low temperatures and the wind. The chief of these are the possible presence of the superoxides and hydrogen peroxides, revealed by the Viking lander biological experiments, and the unfiltered and destructive ultraviolet light that reaches the surface. Both of these factors are highly efficient at attacking and breaking down any organic material and were cited as the cause of the negative GCMS reading from the surface soil. Any carbon-based life forms would have to overcome these problems in order to remain active on the Martian surface.

Endolithic organisms would have some protection against ultraviolet light because of the thin layers of rock under which they live, but, on the assumption that they would need to leave their habitat to propagate themselves in other areas of the planet, some other form of protection would be needed. Various proposals have been made concerning ultraviolet protection methods. For example, Wolf Vishniac[237] proposed that ultraviolet light could be halted by the presence of pigments within the organisms which would fluoresce as a side-effect, thus providing an internal and safe light source for further photosynthesis to take place. It should be noted that no similar mechanism exists in life on Earth, but then there is no need of it here.

The other problem of soil oxides has been dealt with by Earthbound bacteria by using specialised enzymes to break down hydrogen peroxide into oxygen and water. The same mechanisms could be used by life on Mars, with the spare oxygen and water being used for other metabolic processes.[24, 227] Other adaptations proposed for Martian endolithic life include unspecified mechanisms by which the bodies of any organisms could retain moisture and waste products within their cells to minimise wastage, and the possibility that life could actually dwell directly on hydrogen peroxide itself.[227] For these reasons many researchers see endolithic life in

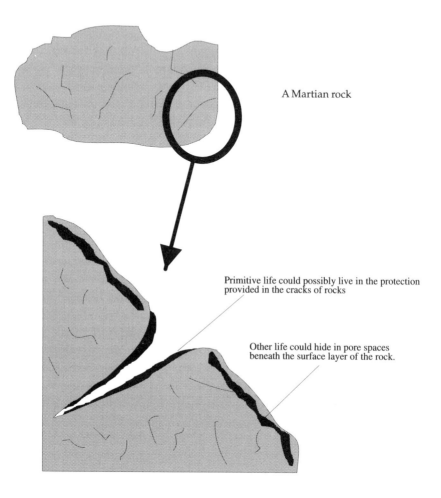

A Martian rock

Primitive life could possibly live in the protection provided in the cracks of rocks

Other life could hide in pore spaces beneath the surface layer of the rock.

Fig. 9. The possible location of endolithic organisms within Martian rocks.

the Antarctic as a good analogy to what life may be like on the surface of Mars, and think that Antarctic life may even be able to adapt to Martian conditions relatively easily. Others violently disagree with this.

The chief objector to the dry valley analogy is Norman Horowitz, designer of the Viking pyrolitic release experiment. He makes a number of highly valid points about the comparisons between endolithic organisms and conditions on Mars.[93, 119] His chief objection is that on Mars the extremely low atmospheric pressure does not allow liquid water to exist on the planet's surface or even within the pores or cracks in rocks; instead, it skips the liquid phase and melts directly from ice into water vapour. The argument that this water vapour could be used by organisms, as occurs in the Antarctic, is also a bad one as the humidity levels on Mars are almost non-existent in comparison to those in the Antarctic, which lies only kilometres from the sea. In fact, says Horowitz, the study of life in the Antarctic may only help to prove that Mars is a sterile planet.

He argues that the fact that 10–15 per cent of Antarctic soil samples can be barren in an environment where water commonly exists in the form of ice and atmospheric humidity demonstrates the complete dependence of life on the availability of liquid water. Although endolithic organisms do survive in the dry valleys, the means by which they do so (warmer summers, plus the availability of liquid water, water vapour and nutrients) are not commonly found on the surface of Mars. Instead, the Red Planet offers a whole host of hostile conditions (oxides, low temperatures, no water, ultraviolet light, etc) to which the endolithic organisms in the Antarctic would have to adapt in order to survive. The theories about how they could adapt are not proved, and require mechanisms about which we can only guess

The endolithic organisms of the Antarctic prove the adaptability of life to our own climatic extremes, but conditions on the surface of Mars make being on the summit of Everest in a gale seem preferable. A bad comparison between the Earth and Mars does not preclude the possibility of life on Mars's surface (see later discussion), but in analogical terms it is unlikely. Better theories about the whereabouts of life on Mars concern the possibility that it lives not on the surface, but under it.

LIFE UNDERGROUND

The Viking landers only sampled the top 10 centimetres of Martian soil in two separate localities, giving us little clue as to what was in the soil layers beneath it. There were also many environments on Mars's surface that the Viking landers did not measure or examine, and since the 1970s onwards a series of new life-bearing environments have been found on Earth that might make better bases for comparison with the surface of Mars than the dry valleys of the Antarctic.

In the early 1970s Carl Sagan proposed that Mars might go through a cycle of cold ice ages and warm interglacials, just as the Earth has over the last 2 million years. This cycle, calculated by Sagan to last 50,000 years, was based on known variations in the tilt of the planet and its orbit through time. Presently Mars is midway between an ice age and an interglacial, and it was theorised that, as the planet warms, carbon dioxide trapped in the polar caps would melt, forming a thicker atmosphere capable of retaining more heat within it. In other words a 'greenhouse effect' would occur.

Sagan proposed that a layer of permafrost (a thick layer of water-saturated soil that becomes frozen beneath a non-frozen top layer of soil) existed over all or part of Mars and that organisms frozen into the permafrost could be revived every 25,000 years during the warmer interglacials. There are problems with this theory, including the later discovery that the polar caps are made of water ice and not carbon dioxide ice, but there are factors about it that still make it plausible.

The first thing to say is that there is some evidence for the existence of ground ice on Mars: probable ice wedge polygons have been seen in the North Polar region of the planet.[223] On Earth these features, which consist of polygonal arrangements of stones forming fields of honeycomb-like structures, are formed by a cycle of freezing and thawing in a permafrost layer which heaves stones to the surface where they collect in the troughs between large ice mounds. The presence of these structures on Mars indicates not only the existence of a layer of permafrost in the Martian soil, but that it must undergo a regular cycle of thawing out and re-freezing. Other authors have also proposed the presence of permafrost near the equatorial regions of Mars.[161, 223]

In terms of the periodic revival of life during the warmer interglacials, there have been many modern examples of dormant ancient bacteria being revived from unusual preserving environments. The oldest reliable case of revived bacteria comes from the centre of a 35 million-year-old piece of amber,[155] although other claims of reviving bacteria from sediments that are 200 million years[88] and even 250 million years[175] old have been made. Bacteria from the Siberian permafrost, thought to be 5 million years old, have also been revived. A 50,000-year revival period, as proposed by Sagan for Mars, does therefore have good terrestrial comparisons and could be a feasible survival mechanism.

Similarly to the permafrost theory, Lynn Rothschild[191] has proposed that organisms might be able to survive above ground inside the polar ice caps themselves. Her theory is based on the discovery of Earth algae and other life that can live in trapped pockets of water within both glacial and sea ice. These organisms survive by shielding themselves from harsh sunlight underneath a protective layer of ice and then by feeding on nutrients blown on to the ice and by fixing carbon from carbon dioxide gas trapped inside the ice.

Although it is possible that such organisms could exist within the Martian polar caps, many of the arguments levelled against the endolithic organisms of the Antarctic can be applied here too. This is especially the case concerning the possibility of trapped pools of water existing within the Martian ice which, with winter temperatures as low as −133°C, would be frozen to quite some depth. Moreover, when the ice caps seasonally melt, the water goes straight into the vapour phase, thus not making any liquid water available on the surface, so that, in terrestrial terms, organisms living in the surface ice on Mars looks to be unlikely.

One further surface habitat for life has also been proposed by Rothschild. This proposal was based on the work of C. Norton and W. Grant, who in 1988 observed that bacteria could become trapped within salt crystals and remain alive for at least six months. This was initially observed in the laboratory, but further research soon found that the same situation occurs in the natural evaporitic basins of the world. Evaporitic basins are areas where a high temperature, dry atmosphere and low rainfall lead to more water being evaporated from an enclosed body of water, such as a lake, than enters it. This leads to a concentration of dissolved minerals in the water so that it becomes salty. As time goes on the salt crystals sink in the lake and accumulate at the bottom, so that when a basin dries out altogether salt deposits many hundreds of metres thick, called evaporites, can be left behind. The modern Dead Sea is a classic example of an evaporitic basin, and it is proposed by Norton and Grant that as the salt crystallises out on the lake bed, it would trap bacteria inside the crystals themselves. Bacteria trapped in modern evaporites could use inorganic supplies of carbon and nitrogen, found in the air, to survive, and it is even suspected that some may have reproduced whilst inside the salt.

Thus Earthly bacteria can survive in evaporites, but is there any evidence that a similar environment could be found on Mars? At both Viking lander sites a thin but solid crust was found on the soil surface that had to be broken to sample the soil beneath. This feature is known as a duricrust; in the Earth's deserts, it is most likely to be formed by salt crystals binding the topsoil together. Other evidence for the existence of salts on Mars's surface was presented by B. Clark and D. van Hart in 1981. This, together with the previous evidence for flowing water, means that in the past all the ingredients necessary for the formation of evaporites on Mars have existed, and it is speculated that many such deposits may have been formed after the disappearance of the surface water approximately 1,500 million years ago.

Rothschild argues that organisms living in Martian evaporites would be protected from ultraviolet light by the salt itself and could gain a supply of water from the salt crystals (up to 7 per cent of salt's weight may be water) and thereafter live by photosynthesis and by the fixation of carbon dioxide and nitrogen gas adsorbed into the evaporite from the atmosphere. She also argues that temperatures within the evaporites could be higher than that of the surface atmosphere.

The entire theory is reliant on evaporites being found on Mars. This is something that has not been proved as yet, but, even if it were, the same arguments can be levelled as at the other specialist environments suggested earlier: just because life has adapted to an environment on Earth, the finding of a similar environment on the surface of Mars does not guarantee that life will be there too—although it does offer hope.

HYDROTHERMAL SYSTEMS

In the 1980s a better idea of Mars's geology was gained from the Viking orbiter photographs. Features on these photographs, including the dry rivers and volcanoes, gave hope that, beneath the surface, heat from the planet's interior would allow warm liquid to exist in the fractured, porous rocks of the crust. Where there is liquid water there is the hope of life, and so it was that the search for life moved from the surface of Mars to deeper within the planet's crust.

In addition to the ancient dry river beds seen on the older parts of Mars's surface, other scientists found newer, less eroded river channels running across much younger lava flows. Some of these newer channels were connected with the volcanoes on Mars, suggesting that the heat associated with the volcanoes was capable of melting frozen water trapped under the surface. The width and depth of some of these canyons suggest that a considerable amount of water must be locked up within the crust. Meteorites from Mars (see Chapter 11) indicate that volcanic activity was still working on the planet until at least 200 million years ago. In terms of the age of the planets (5,000 million years) this is very recent, and whilst no active volcanoes can currently be seen, it remains likely that the interior of Mars is still hot and partially molten. If so, then it is possible that a region of geothermally heated liquid water exists between the frozen surface and the superheated planetary core. This liquid layer would exist in the pores of sedimentary rocks or in the cracks and fractures of igneous rocks.

Systems like this exist all over the modern Earth, and in particularly geologically active areas, such as Yellowstone Park in America, the heated layers of water are in contact with the surface, where they form bubbling mud baths, geysers and hot springs. On the sea bed, too, volcanically heated water, rich in dissolved minerals, pours out through cracks in the ocean floor to form 'black smoker' hydrothermal vents, which were filmed for the first time only in the mid-1980s.

On Earth viable communities of animals and plants are associated with both terrestrial hot springs and the oceanic black smokers. Both of these are considered extreme environments for life to survive, and, naturally, analogies have been made with Mars and the possibility of hydrothermally heated water existing there. However, as with the dry Antarctic valleys, these analogies are not particularly good ones. Both environments have access to freshly aerated (i.e. oxygenated) water combined with the presence of light (in the case of hydrothermal springs) or nutrient-giving sulphides (in the case of deep sea vents). Here there is a fundamental difference between the sorts of hydrothermal water cycle that could exist on Mars and those on Earth.

Although hydrothermal systems on both planets would be heated from their interior, the water in the Earth's crust would be derived from the planet's surface whilst that on Mars could not be as no water exists there. The water on Earth is therefore being constantly re-oxygenated by its contact with the atmosphere, and any organisms living in it would benefit from this. Water coming near the surface of Mars would evaporate instantly (or remained trapped below a frozen layer of ground), and therefore any water currently residing within the Martian crust must have been isolated there for hundreds of millions of years. This would not necessarily make the water stale, as volcanic activity beneath the hydrothermal layer would produce fresh water supplies along with nutrient minerals (such as iron, sulphur and copper) and gases (such as carbon dioxide, sulphur dioxide and methane) which would be pumped into the older water supplies to revitalise them.

This environment, with its heat, volcanism and lack of oxygen and light, might sound hostile to life, but on Earth there is a group of bacteria that would absolutely thrive on it. They are collectively known as anaerobic autotrophic organisms (AAOs), or chemolithautotrophs, which simply means that they need neither oxygen nor sunlight in order to survive. Instead these bacteria utilise the carbon atoms in carbon dioxide (and other gases) to build up their bodies and respire using other gases such as carbon monoxide. Many groups of these bacteria, including methanogenic and acetogenic bacteria, do not require organic matter in order to grow,

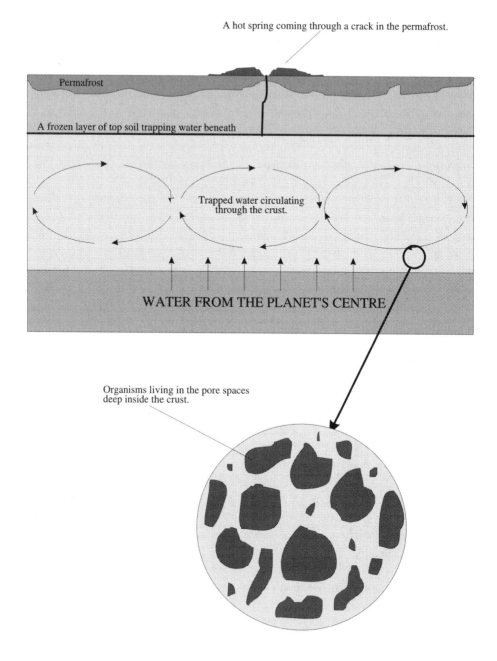

Fig. 10. The features of a hydrothermal system
deep within the Martian crust.

merely the presence of water containing dissolved volcanic nutrients and gases.[23]

On Earth the range of hostile environments where AAOs have been found is truly amazing. On the surface they are regularly found in anaerobic hydrothermal springs, where they can comfortably survive temperatures of 70°C or more.[215] They have also been found in volcanically heated soils[98] where temperatures are between 80 and 110°C, but their most spectacular occurrence has been in the superheated briny water that is associated with oil wells and aquifers that may be buried up to 7 kilometres deep within the crust.[202] Here we see a direct comparison between an inhabited Earth environment and one that could very well exist on Mars.

Other research that has been done on AAOs also gives encouragement to the possibility of their counterparts existing on Mars. Genetic work carried out on a wide range of plants, animals and microbes has suggested that a member of the AAOs, a sulphur metabolising bacterium that lives in hot springs, was a common ancestor to all the life on Earth.[123, 245] This suggests that the first organisms on Earth were the AAOs, a theory that makes sense when one thinks of the carbon dioxide-dominated atmosphere and higher instance of volcanic activity in the past.

If AAOs were also the first type of organisms to evolve on Mars then, as the water disappeared from the surface, they could have continued to survive in underground hydrothermal systems, and there is strong evidence that these systems still exist under the Martian surface. Huge flood channels exist on the surface, some of which are associated with the recent volcanic activity on Mars. This suggests that the volcanoes have melted the permafrost layer, allowing the meltwater, and any stored underground water, to cascade across the planets' surface. This is strong evidence that water is indeed present in the Martian crust and that deep down it is probably liquid.

Other people claim to have found evidence of actual hot springs on Mars based on the Viking orbiter photographs, but better confirmation can be expected from the higher-resolution camera on the Observer orbiter in 1998. Until recently the possibility of surface hot springs on Mars was thought to be unlikely because of the low atmospheric pressure and thickness of permafrost that they would have to penetrate to reach the surface. However, in 1997 a NASA researcher, Virginia Gulick, proved that hydrothermal systems could theoretically exist on the current Martian surface, and in the same year a spring was found in the Arctic that had come through 500 metres of permafrost.[141] In comparison to the other theories discussed in this chapter, the possibility for AAOs in hydrothermal systems on Mars is a good one as the organisms have the capability to have evolved and survived there, and there is evidence that the hydrothermal systems do still operate on the planet. Furthermore, the ALH84001 meteorite may also contain fossils from just such a hydrothermal system (see Chapter 11).

WHERE TO SEARCH FOR LIFE

Given the types of environments outlined above, and using the information gathered from the Viking orbiter and other Martian probes, it is possible to suggest the best places on the planet to search for the various forms of life discussed above.

Ice-dwelling biota would most likely be found in either the polar ice caps or the permafrost which is thought to be restricted to above 45° north and south of the equator. It is also possible that permafrost would be found in the beds of the ancient rivers and canyons, where water could have soaked into the ground during Mars's wet phase and then frozen. The same is true of evaporitic rocks, which would also most likely be formed after the evaporation of water from within the river beds, possibly leaving behind salt deposits in which organisms could survive. Evaporites are best formed in slightly hotter climates, so it may be more productive to look for evaporitic deposits in the dry river beds nearer the equator.

Endolithic organisms, like those in Antarctica, could theoretically occur anywhere on the planet where there are surface rocks. The warmer climate nearer the equator may encourage growth in organisms, but, equally, nearer the poles there is more ice and therefore more water and water vapour available. Again, frozen water may exist in the bases of the dry rivers and lakes, and the atmospheric pressure is slightly higher inside the deepest basins and canyons, increasing the possibility of water remaining liquid.

The underground hydrothermal water systems would only be found some hundreds of metres or even kilometres below the planet's surface and would therefore be hidden from our gaze. Even so, they are liable to be associated with volcanism, and therefore a search would best be concentrated on the younger volcanoes such as Olympus Mons. The only external evidence of their presence would be small, hydrothermal vents on Mars's surface, as with geysers and hot springs on Earth. Hydrothermal springs on Earth are excellent places for fossils to be preserved as the bodies of organisms become incorporated into mineral deposition around the springs. On Mars any organisms living in the springs, or brought to the surface from deeper in the crust, would be likely to be preserved, making them excellent places to look for evidence of life.

Summing up, therefore, it would appear that the best places to search for organisms would be in the dry river beds of Mars or in any hydrothermal springs that may be identified. Apart from endolithic organisms, any new search for Martian organisms will need a probe that can drill into the surface of Mars (or its ice caps) to obtain samples from some way beneath the soil. In the case of evaporites and permafrost, drilling for only a few metres may be all that is necessary to find suitable samples. With the deeper hydrothermal systems, crustal cores would need to be taken, and these may need to be several kilometres in depth. To commission a similar drilling operation on Earth would require much technical expertise and money, making such a venture on Mars unlikely in the near future. A possible alternative to this would be to try to find the aforementioned hydrothermal outlets nearer the surface, such as hot springs, from which samples could be taken, although an absence of organisms in this environment might not preclude their existence deeper in the crust.

Endolithic organisms could be searched for in the same manner as the Pathfinder Sojourner has investigated the geology of rocks by pressing a measuring instrument against the surface of a rock. A sample return mission, possibly planned for 2003, would bring back rocks that could be examined in minute detail for signs of life.

It is therefore clear that, apart from the endolithic option, a mechanism would be needed to drill or penetrate some distance into the surface of Mars. The Russians tried to overcome this problem by attaching two wedge-shaped probes to their Mars 8 mission. These would have dropped through the Martian atmosphere and then buried themselves up to 2–3 metres into the soil, where they would then have made measurements and carried out chemical analyses on the soil. Unfortunately, the Mars 8 spacecraft was destroyed shortly after its launch. However, NASA are looking at including similar sub-soil devices in future missions. Other probes could be designed to drill short distances into the crust or polar caps, but the energy needed to do this would not make it an entirely practical method of exploration or indeed the best use of a probe on Mars. Realistically, until manned exploration is a possibility, we are going to be limited to sampling only what lies on or just below the surface of Mars.

PART THREE

FOSSIL LIFE ON MARS

10. ANCIENT LIFE ON MARS

THE QUESTION OF FOSSIL LIFE

The realisation that life was not abundant on the modern surface of Mars led researchers to ask the next inevitable question. If we do not find life on Mars today, could it have existed there in the past?

In the late 1960s the answer to this looked entirely negative as the first Mariner photographs of the Martian surface revealed a cold Moon-like planet unlikely to have changed since its initial formation. The arrival of the Mariner 9 photographs provided a different perspective on the issue as a host of geomorphic features, such as dry river beds and volcanoes, appeared, showing that the planet was not as inactive as was first thought (see Chapter 6). Later the Viking orbiters mapped the whole of Mars in high resolution, allowing these features to be studied in detail so that geologists could reconstruct the evolution of the planet and the landforms upon it. Mixed in with this geological reconstruction was still the question of whether life had ever played a role in the history of Mars. During the years of study that have taken place since the Viking mission this question has been addressed many times and the answers have been surprisingly positive.

EARLY MARS AND WATER

The reason for the change of mood in exobiologists after the arrival of the Mariner 9 photographs was because the presence of rivers, canyons and outflow channels on Mars proved that water had at some point existed on its surface. As we have discussed previously, a prerequisite for life on other planets is the prolonged existence of liquid water on its surface so that the necessary chemical reactions could have occurred that would lead up to the origin and evolution of life, or, in the panspermia theory (see Chapter 13), so that incoming extraterrestrial life could have found a suitable environment in which to breed. The presence of water also suggests that the atmosphere on Mars must have been thicker than it is currently, where any liquid water on the surface boils at 0°C. A thicker atmosphere might mean more humidity, clouds, rain and even some form of ozone layer to protect any life from the harsh ultraviolet light that currently strikes the surface, breaking any organic molecules apart. The prospects for life having evolved on Mars in the past looked very promising.

For life to have arisen it would have needed a considerable period of time in which the liquid water was stable on the planet's surface. On the first examination of the Mariner 9 photographs this did not look very hopeful as the majority of the water-associated features were restricted to the older cratered areas of Mars and not the younger volcanic regions, indicating that water had only been present during the meteorite bombardment of 3,800 million years ago. In addition to this, most of the features seen were outflow channels, which are large-scale landforms associated with violent floods on Earth. On Mars each outflow channel begins suddenly from a depression in the surface and then cuts a single channel, up to 200 kilometres wide, across the Martian plains until it simply fans out into the desert and stops. In terms of life this was bad news as it indicated that the water-carved features on Mars were the product of sudden and catastrophic floods that erupted from underground and then flowed away to be soaked into the desert or evaporated, leaving little, if any, time for life to evolve in their short, violent existence. The outflow channels did at least prove that there must have been a considerable amount of frozen water hidden within Mars's crust and gave hope that it might still be there. The higher resolution of the Viking photographs revealed features that gave more encouragement to the exobiology fraternity.

In addition to the violent outflow channels, more delicate and permanent water erosional features were found to indicate that water may indeed have had a long-term existence on the planet's surface. Again, within the cratered older regions of Mars there were discovered a whole series of much narrower, branching channel networks that very strongly resemble river drainage systems on Earth. These rivers are only 1–3 kilometres wide, with flat floors, steep walls and an increase in size downstream, and have tributaries,[39, 242] indicating that they were formed by the long-term erosional action of slow-flowing water rather than a sudden burst of floodwater. A number of these river networks were found to be associated with the younger volcanic regions, indicating that they were active up to about 200 million years ago. A detailed study of the river networks showed that many had their origins in the highland regions of Mars, posing the question where the water was initially coming from in order to feed them.

With the outflow channels, it was easy to imagine a meteorite impact or volcanic activity suddenly melting vast volumes of frozen groundwater to form a flood. However, the longer-term river valleys needed a more sustained and gradual source for their water. The presence of numerous river channels on the highlands suggested that perhaps the water was arriving in the form of rain, which would then soak into the ground, form springs and drain down into the lowland regions as it does on Earth.[222]

For it to rain on Mars the atmosphere would have to have been considerably thicker, wetter and warmer than it currently is, all of which favours the origins of life. This view of a warmer, wetter early Mars was favoured during the 1980s, the speculation being that an increased amount of carbon dioxide in the atmosphere would lead to water vapour and heat being trapped and that a greenhouse-style heating of the planet's atmosphere would occur, allowing the stable formation of clouds and rain. In recent years this vision has been scaled down slightly, and it is no longer thought probable that the atmosphere would have been thick or wet enough to sustain precipitation in any form; the short length and low drainage density of the river channels has been used as evidence for this.[222] Instead, it has been proposed that the rivers' source may have been hot springs or the slow melting of subsurface ice. Whatever their origin, the rivers did exist and are prima facie evidence of the durability of liquid water on Mars, and it is this that is crucial for life and not necessarily a warmer climate. It should be added that other authors have suggested that the rivers may have been formed, not by water, but by geological faulting,[210] erosion by lava flows,[38, 208] ice,[132] wind[53] and liquid carbon dioxide,[201] although all of these are thought to be unlikely.

The presence of liquid water in rivers is all very well, but it would still be difficult for life to have evolved in a flowing body of water which was only 100 kilometres or so in length. Most theories about the origin of life on Earth require large, stable accumulations of water, in the form of seas or oceans, where the necessary reactions could occur without periodic disturbance by changes in stream flow or water supply. It seemed logical that the valley networks on Mars must drain somewhere, and it was hoped that they would drain into bodies of water rather than simply ending in the desert like the outflow channels do.

The first evidence for large bodies of water having existed on Mars was found in the Mariner 9 images which showed that the Valles Marineris region apparently contained layered sedimentary deposits likely to have been laid down underwater. The Valles Marineris is a series of wide, deep canyons that run parallel to each other across a 2,500-kilometre swath of central Mars. The layered deposits were seen inside these canyons and indicate the periodic accumulation of material brought into the canyon by the action of water, wind or volcanoes or simply through the sides of the canyon falling down.[154] The estimated depth of these sediments is up to 5

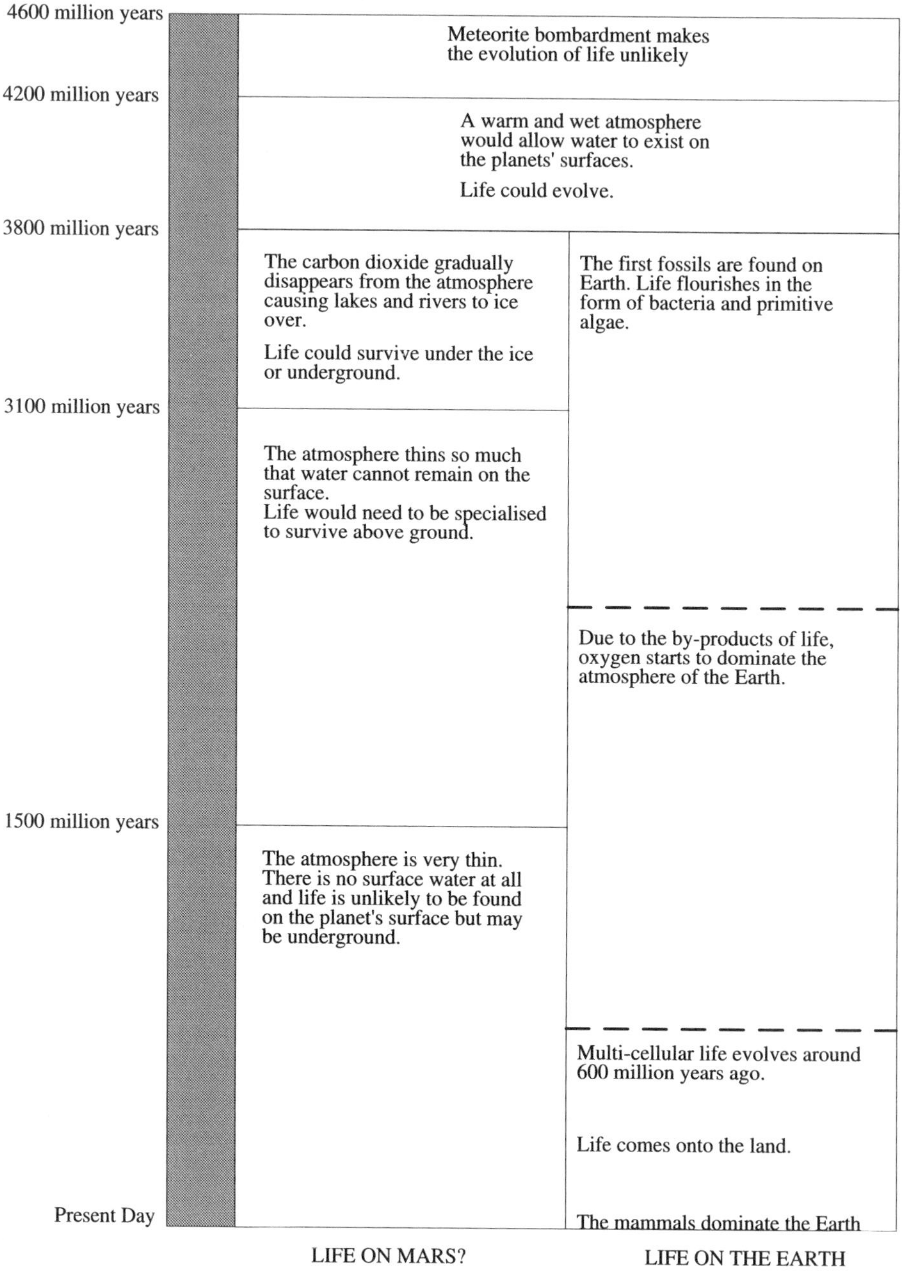

	LIFE ON MARS?	LIFE ON THE EARTH
4600 million years	Meteorite bombardment makes the evolution of life unlikely	
4200 million years	A warm and wet atmosphere would allow water to exist on the planets' surfaces. Life could evolve.	
3800 million years	The carbon dioxide gradually disappears from the atmosphere causing lakes and rivers to ice over. Life could survive under the ice or underground.	The first fossils are found on Earth. Life flourishes in the form of bacteria and primitive algae.
3100 million years	The atmosphere thins so much that water cannot remain on the surface. Life would need to be specialised to survive above ground.	
		Due to the by-products of life, oxygen starts to dominate the atmosphere of the Earth.
1500 million years	The atmosphere is very thin. There is no surface water at all and life is unlikely to be found on the planet's surface but may be underground.	
		Multi-cellular life evolves around 600 million years ago.
		Life comes onto the land.
Present Day		The mammals dominate the Earth

Fig. 11. A brief outline of the history of life on Mars versus life on Earth.

kilometres and their lateral continuity in the canyons led scientists to conclude that they were likely to have been deposited in a standing body of
water.[154] The scientists had found a possible site for the evolution of life.

A study of the Viking photographs found that the Valles Marineris was
not the only place on Mars to have had accumulations of water (which
were termed 'palaeolakes'). One of these areas, the Elysium Basin, has been
studied in particular detail and has been used as a model to show how life
could have evolved and continued to live on early Mars. The Elysium Basin is a large regional feature that stretches for 2,500 kilometres along the
northern edge of the equator. It is an elongated canyon that is 2,500 metres
deep in its centre and is tens of kilometres across at its widest part. Water
appears to have flowed into the basin through a number of valley networks, and there are also a number of areas where water may have flowed
out of the basin.[212] An old shoreline has been identified in the canyon that
would have made the palaeolake over 1,000 metres deep at its centre. This
was clearly a considerable body of water that must have remained stable
for some time on the planet's surface, and, after the discovery of a further
fifteen similar palaeolake regions,[242] the vision of early Mars moved from
one of a desert-like planet to one with a thicker climate and drainage networks that ran into large bodies of water which covered up to one-seventh
of the planet's surface. All very well—but what are the chances of life evolving under such conditions?

LIFE IN THE LAKES

Most of the work on the prospect of life in Mars's palaeolakes has been
done by NASA's exobiology unit and in particular by Christopher McKay.
Using the geological reconstruction of the early Martian atmosphere and
analogies between the Earth and Mars, McKay and others have proposed
four main epochs in the history of the evolution of Martian life.[143, 242]

The first epoch is the period in Mars's history when it and the Earth
were most similar. It begins approximately 4,200 million years ago, when
the planet's crust hardened and large amounts of volcanic activity pushed
out carbon dioxide and other gases from the molten core to create a thick
atmosphere. It is felt likely that this atmosphere would have been thick
and warm enough to support widespread liquid water on the planet's surface. It would be during this time that any life would be likely to originate
in the vast palaeolakes that would have covered Mars's surface.

In the second epoch, this era of a warm and wet atmosphere began to
decline 3,800 million years ago as the carbon dioxide was slowly stripped
from the atmosphere through its incorporation into carbonate sediments
and its escape into space. With little volcanic activity to revitalise the car-

bon dioxide levels in it, the atmosphere would have become thinner and the temperature lower. It is felt certain that the temperature would have dropped below 0°C and may have been as low as –40 to –70°C, leading to the freezing over of the palaeolakes. Although the temperature outside may have been hostile to life, it is speculated that, beneath a protective layer of frozen ice, conditions within the palaeolakes may have been very suitable for life and it could have continued to thrive.

In 1985 Christopher McKay studied a series of ice-covered lakes in Antarctica and found that the layer of ice on their surface acted as a means of insulating the water below from the freezing atmosphere. Heat within these lakes came from any sunlight penetrating the ice layer, geothermal heat from below and the energy released from water freezing on the underside of the ice. The only mechanism for heat loss would be through the ice layer itself which, at –40°C, was 19 metres thick.[242] Nutrients, gases and sediments come either through the ice or under its edges. This model was applied to the conditions in the second epoch of Mars, where it was postulated that the equable conditions under these ice-covered lakes could have preserved a suitable environment for life for up to another 700 million years.

This brings us to the third epoch, which has been estimated to have occurred from 3,100 to 1,500 million years ago. During this time the loss of carbon dioxide was so severe that the thin atmosphere was no longer able to support free-standing surface water at all and the lakes would have slowly evaporated away or possibly sunk into the ground beneath them. In the latter case, liquid water could possibly have continued to survive in the pore spaces and cracks within the sedimentary rocks of the river and palaeolake basins, and so here, too, life could have evolved to survive, living in the same manner as the Antarctic endolithic organisms discussed in Chapter 9. As environmental conditions worsened and the planet moved into the fourth and final epoch that it is in now, life would not have been able to survive on the surface at all and would be forced to separate itself from the atmosphere in order to find water to survive. The possibilities for these final two epochs are covered fully in Chapter 9.

Even with the availability of large bodies of liquid water on Mars, would life have been likely to have evolved there at all, or are the processes that created life on Earth truly unique in this solar system? To answer this we have to look to see under what conditions life on Earth evolved and how quickly it managed to do so.

THE ORIGIN OF LIFE ON MARS

Current thinking has the Earth and Mars sharing near-identical environments until about 3,800 million years ago, when the paths of the two planets separate, Mars's atmosphere rapidly becoming hostile to life whilst Earth's remained equable. It is thought by most exobiologists that if life had not evolved on Mars by 3,800 million years ago then it is unlikely to have done so at all.

There is a problem with this time limit in that all the planets in the solar system underwent an intensive period of meteorite bombardment from their creation through to approximately 4,000–3,800 million years ago. It was this period of bombardment that cratered the Moon and the older surfaces of Mars. Aside from the physical damage it caused to the planets, it is thought that life would not have been able to originate and thrive for any length of time before being hit and destroyed by a meteorite impact. It is therefore generally felt that the earliest opportunity for life to evolve in the solar system was 4,000 million years ago and may have actually been later than that.[135] But how long would it take for viable life to evolve?

The oldest positively identified fossils on Earth are 3,500 million years old and show that life was advanced enough by then to form colonial mats of layered algae, called stromatolites, suggesting that it had been established for some time prior to this.[209] However, there is strong geochemical evidence that life may have been up and running in the oldest known sedimentary rocks on Earth, which are from the 3,850 million-year-old Isua region of Greenland.[207] If these rocks do contain evidence of life, then it took a very short space of time indeed—less than 250 million years—for it to originate and become established on Earth after the end of the meteor bombardment 4,000 million years ago. This gives hope that a similarly rapid evolution could have occurred on Mars before its palaeolakes froze over approximately 3,800 million years ago. Even if the Isua rocks do not contain evidence of life, the 3,500 million-year-old stromatolites show that it took less than 500 million years for life to occur on Earth, again offering hope that life may have been able to evolve quickly enough on Mars to allow it to become established whilst liquid water still remained unfrozen. However, even with liquid water, other conditions may have needed to be met before life could have arisen on Mars.

I do not wish here to enter into the subject of the method by which life originated on Earth (or on any other planet) as this is dealt with in Chapter 13. Most of the current theories require liquid water as a medium in which the various chemical processes necessary for life to begin could occur. Recent genetic work on ribosomal RNA from a wide range of organisms suggests that the common ancestor to all life on Earth may have been a bacterium that lived in the superheated waters surrounding the sulphur-rich black smokers located on the deep sea floor.[245] Such an environment would have almost certainly been available on early Mars, whose crust would have been fractured and split by volcanism and meteorite impacts, allowing water to seep into it to become heated and later expelled when full of dissolved minerals. It is even possible that black smokers could have survived beneath the ice-covered lakes, and that their equivalents may even exist below the Martian surface today (Chapter 9). It has also been suggested that a wide variety of completely different life forms evolved on Earth but that it was only the heat-loving ones that were able to survive the period of meteorite bombardment to become our ancestors.[218] Perhaps the same was true on Mars.

Other than water, it is thought that the presence of nitrogen, in its fixed nutrient form of nitrate, would also be a useful precursor to life as it forms a vital part in the formation of amino acids, RNA and DNA on Earth. Fortunately, nitrogen gas would have been available on early Mars from the gases associated with volcanic activity or meteoritic bombardment and could have been fixed into nitrates initially from lightning strikes but perhaps later from biological fixation.[143] Other theories concerning the evolution of life involve the presence of clay as a supporting structure upon which life could evolve,[30] or bubbles within which chemicals could be brought together.[55] These would have been available on early Mars as well, making its similarities to early Earth very close indeed. Although it cannot be certain, the similarities between the two planets does make the evolution of life on Mars seem as likely as that on Earth. If, then, we ever make it to Mars ourselves, what types of fossil evidence should we look for, and where should we look?

WHERE TO LOOK . . . ?

On Earth the main factor governing the distribution and state of preservation of fossils is the rock type in which they are found. Fossils are almost exclusively found in sedimentary rocks which have been laid down on land or underwater. The accumulation of sediments, such as mud, sand, peat and limestone, allows the dead bodies of any organisms living on or near them to become buried inside, thus protecting them from rotting and the wear and tear of the external environment. In effect, the process of fossilisation is akin to conserving a flower by keeping it pressed between the pages of a book. In exceptional circumstances, fossils can be preserved in rock types other than sedimentary ones, such as inside the cracks and pores of volcanic rocks, and although these are rare on Earth, they may play an important part on Mars and are one of the key issues regarding the ALH84001 meteorite (Chapter 11).

Once an organism has become fossilised, that is not the end of the story: in fact, what happens to the fossil after it has been preserved is more im-

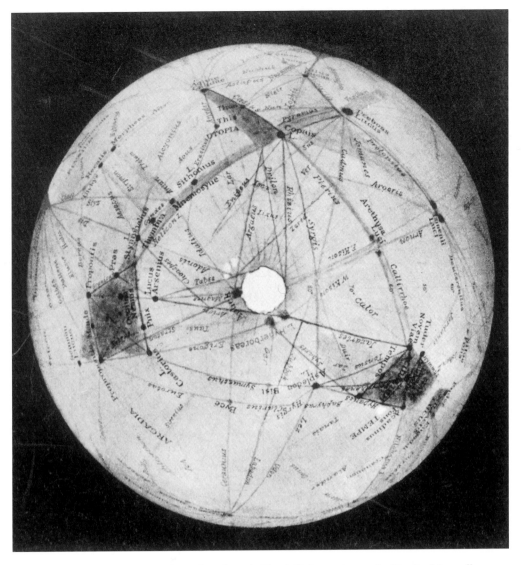

A view of Mars and its canals from the planet's North Pole, as drawn by Percival Lowell. Clearly visible are oases, areolas and geminated canals. (Royal Astronomical Society)

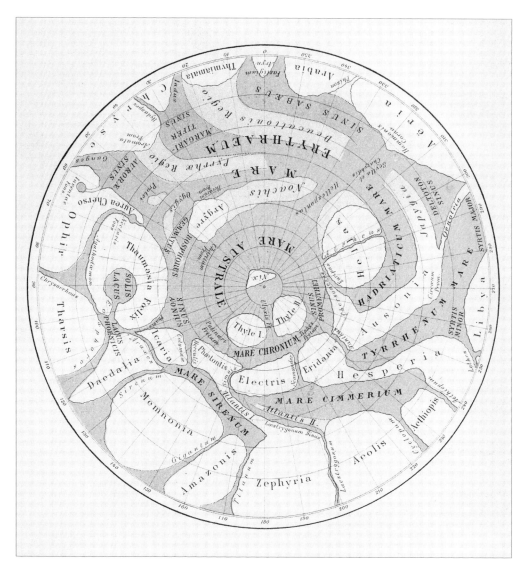

Mars's North Pole as drawn by Giovanni Schiaparelli in 1877. His drawing reflects the belief that Mars had oceans on it. The dark, narrow channels on this map were termed *canali* and would later be described as canals. (Royal Astronomical Society)

Right, upper: The first photographs of the Martian surface by Mariner 4 revealed a cratered landscape like that of the Moon. The possibility of life seemed remote. (NASA)
Right, lower: After the disappointment of previous missions, Mariner 9's photographs showed evidence that liquid water may once have flowed on the surface of Mars. Where there is water, there may be life. (NASA)

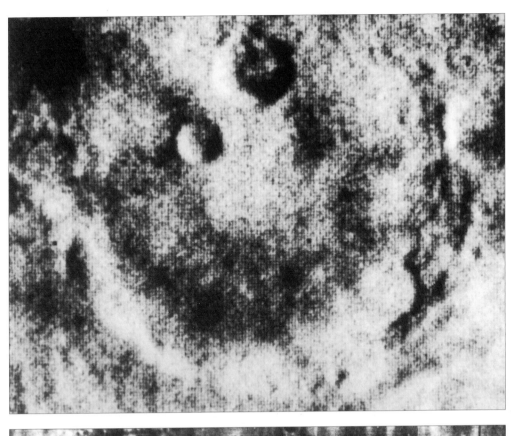

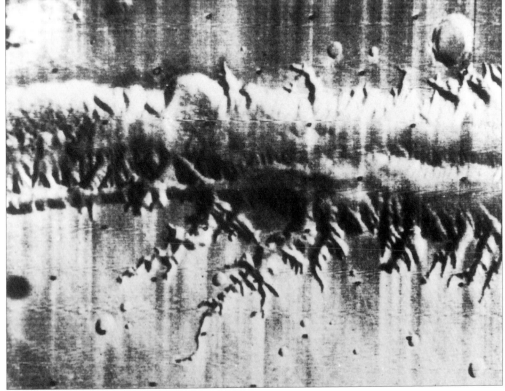

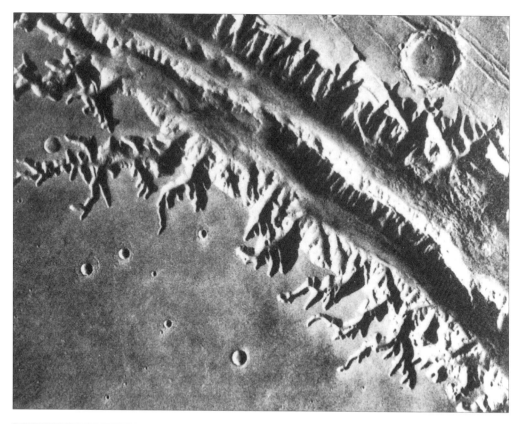

Above: The 300km wide Isu Chasma canyon on Mars. This is thought once to have been the site of a vast lake which would have been a suitable place for life to evolve. Notice the improvement in photographic quality between this Viking image and that taken by Mariner 9. (NASA)

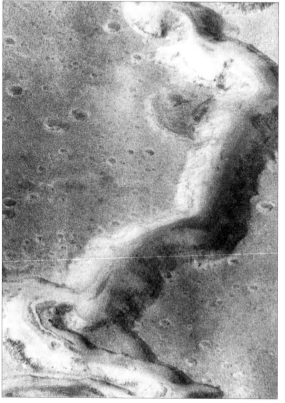

Left: The launch of the Mars Surveyor marked a new era in the exploration of Mars. This photograph of a dry river valley was taken when the probe was still a year from its closest orbit around the planet, March 1998. (NASA)

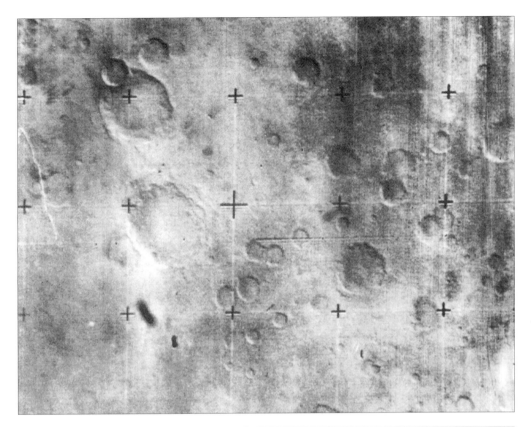

Above: A view of Mars taken from one of the few successful—but still only partially successful—Russian Mars probes, Mars 4. (NASA)

20 KM

Right: A photograph of a dry river bed taken from a Viking orbiter shows clear evidence that water once flowed on the surface of Mars—but was it there long enough to stimulate the evolution of life? (NASA)

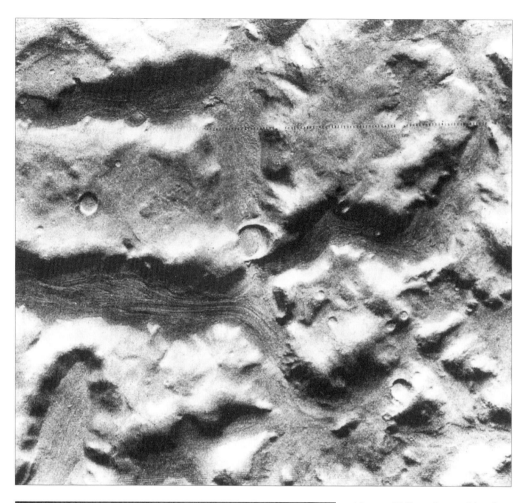

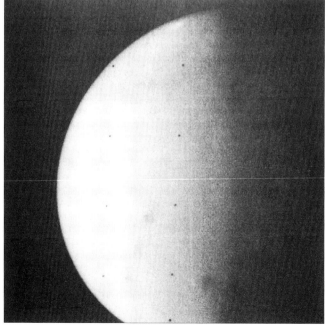

Above: Valleys formed by the presence of ice in the soil show that there is an active permafrost layer in the Martian soil. Some have proposed that life could live in or underneath the permafrost. Viking Mosaic 211-5207. (NASA)

Left: A view of Mars as the Mariner 9 probe approached the planet shows the surface almost completely occluded by a massive, planet-wide dust storm. Such storms were, in the nineteenth century, mistaken for vegetation growth. (NASA)

Above: This view of the surface of Mars was taken in July 1997 by the Pathfinder mission. The mission was later to confirm that Mars had indeed been a wetter and more geologically active planet than it is now. (NASA)
Below: A photograph of Martian meteorite ALH84001, in which it was claimed that there were signs of fossilised life from Mars. (NASA)

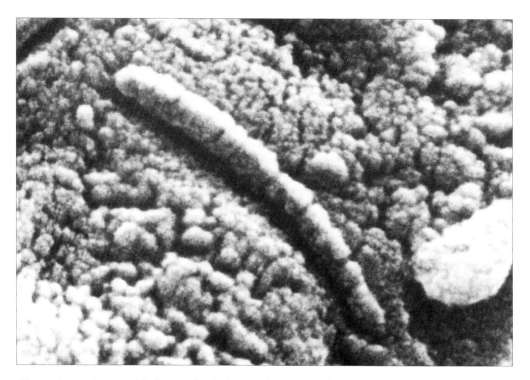

Above: At one-hundredth the width of a human hair, could this feature within Martian meteorite ALH84001 be fossilised life from Mars? (NASA)
Below: This face-like feature on Mars's Cydonia Plain, which is approximately 1.5km in length, has been cited as evidence that extraterrestrial beings have built monuments on the surface of Mars. Viking frame no 35A72. (NASA)

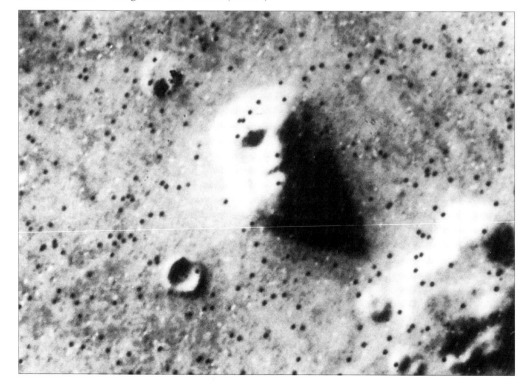

portant than its burial in the first place. On Earth, the number of fossils available for study decreases exponentially with increasing age. The older the rock, the less likely we are to find fossils in it. This is why the fossils of mammoths, which died out only 10,000 years ago, are so common (including frozen specimens), whereas dinosaurs, which were far more numerous for a much longer period of time than the mammoths, are a much rarer find as they died out 65 million years ago. The reason for the disappearance of fossils through time is that once an organism becomes fossilised, that fossil then has to survive the pressure and temperatures that occur as further sediments are piled on top of it. The pressures may crush the fossil, whilst an increased temperature and the hot fluids that percolate through most sediments may dissolve the fossil altogether. If the rock gets buried too deeply, then it may partially or fully melt so that it becomes part of the Earth's liquid mantle once more, thus destroying any evidence of life that it may have contained. Fossils are rare in rock over 600 million years old on Earth.

As was discussed previously, there is strong evidence that large accumulations of sediments may have been deposited in the palaeolakes and river valleys found on the Martian surface. It is to these areas that we should go in search of fossil life should we ever visit the surface of Mars. Indeed, both the Viking and Pathfinder missions were targeted at river valley locations in the hope of finding life and, in the case of Pathfinder, because the sediments in these areas would contain many different rock types washed downstream. In many respects the sediments on Mars may in fact be a better place to look for fossils than those on our own planet. The Earth is a very geologically active planet with a crust that is both thin and very pliable. In certain areas the weight of sediments on the crust can cause it to sink, burying them deeper and bringing them closer to the hot mantle.

Although Mars has been geologically active in the past, its crust is much thicker than the Earth's and consequently any sediment accumulations are unlikely to have dented the crust beneath them. Instead they will just sit on it, where they will either accumulate in hollows, such as canyons or the lakes, or be moved on by further wind or water movement. The net effect of this will be that the sediments will be protected from the heat of the mantle below, increasing the chances of their fossils being preserved. Other processes that are hostile to fossils may still occur, such as the pressure of overlying sediments and dissolving waters percolating through the rocks, but the removal of heat from these processes greatly reduces their effects on the fossils. In addition, the disappearance of water from Mars's surface, probably between 3,100 and 1,500 million years ago, means that any sediments have merely sat on the planet's surface since then with little or, more probably, no water running through them to cause alterations of any kind. In fact, many of these sediments may have been frozen by the harsh temperatures on Mars, dramatically slowing the rate at which any chemical reactions can occur, including the degradation of organic matter. With these factors in mind, many scientists think that the sediments of Mars, particularly those in the palaeolakes, may hold better evidence of the origin of life in the solar system than does Earth.[141]

The palaeolakes are likely to be the best places to search for fossils for a number of reasons. The first is that these lakes were deposited in deep depressions on Mars's surface and therefore will contain thicker piles of sediments than the river valleys, which merely scooted their way across the surface of the planet. Secondly, the palaeolakes would have held large bodies of water, which are better environments for life to evolve, flourish and breed than a fast-moving water channel. The proposed ice-covered lake model also means that they would have been a more hospitable place for life on Mars for millions of years after the climate began to degrade. Thirdly, when life died in the lake it would have fallen to the lake bed, where it could have been quickly covered by incoming sediments to preserve its remains. In rivers any dead bodies are likely to be swept along by the flowing water and they will probably get destroyed by the turbulence— and are certainly unlikely to be covered by sediment unless it got swept into a quiet pool or lake. Finally, the still waters of the lakes would be better places for finer sedimentary particles, such as clay, to settle, where they may cover an organism better and more fully than coarser, sand-sized particles, again increasing their chances of preservation. It has also been suspected by scientists studying Mars[142] that some chemically precipitated rocks, such as limestone (carbonates), which are suitable for fossil preservation may have been laid down within the lakes in large quantities, although this has yet to be fully proved. The palaeolake sites are therefore the place to begin our search for fossils . . . but what exactly are we looking for?

. . . AND WHAT TO LOOK FOR?

This is the sixty-four thousand dollar question and is another one that can currently be answered only in respect of our knowledge of Earth. The search for ancient life on Earth is a relatively straightforward one. Modern genetics has all but proved that every organism on our planet is related one to the other, and so by searching the fossil record we are really only looking for variations on life that already exists on the planet. This makes it much easier to know where to look for fossils and to recognise them when they are found. We do not know what to expect on Mars in terms of what any past life may have been like, but, based on the Earth's fossil record, we can take a guess.

The first thing to note is that any past life on Mars is liable to have been primitive, structurally simple and probably similar to modern bacteria on Earth. This is because life on Mars is liable to have either become extinct or been driven underground before 1,500 million years ago, when the last of the surface water disappeared. The first signs of complex multicellular life on Earth did not emerge until 700 million years ago and did not become common until 570 million years ago. Therefore life on Earth is thought to have been around for over 3,000 million years before it made the evolutionary jump from simple bacteria-like organisms to the skeleton-building multicellular animals and plants such as we see today. By 1,500 million years ago, when the Martian surface water had gone, life on Earth was still 750 million years away from its greatest evolutionary step. Up to and beyond this time life on Earth would have enjoyed equable climatic conditions and abundant water; life on Mars would not, and, apart from the first few hundred million years of the planet's history, would have had to struggle against the odds. Based on the terrestrial experience, it is therefore felt improbable that, given the time involved, life on the Martian surface could have evolved beyond the simple single-celled organism stage (see Chapter 13 for a fuller discussion of similarities between Earth and Martian life). If life retreated below the Martian surface in rock spaces or fractures, then it is unlikely to have had the room to expand into anything bigger than bacteria-sized life. This presents any future palaeontologists on Mars with a problem in their search for fossils.

The fossils that we are all familiar with in museum displays and textbooks, such as dinosaurs, sea shells, ammonites and bones, all represent the hard parts of an animal or plant, with the soft tissue having long since broken down and not being preserved. As it is the hard parts of organisms that are preferentially preserved, any organisms without hard parts inside their structure, such as jellyfish and sea anemones, do not get preserved save under the most exceptional of circumstances. As the majority of the fossil record, before 500 million years ago, consists of organisms that have no hard parts whatsoever, there are huge gaps in our knowledge of life. It is estimated that since the evolution of organisms with skeletons 500 million years ago we have described some 250,000 species from the fossil record, but that during this time there are liable to have been 4,500,000 species. In other words, 95 per cent of the animal and plant species that have lived on Earth during the last 570 million years are unknown to science.[173] On Mars it is expected that life would not have got beyond the bacteria-like stage, which, the life forms being entirely soft-bodied, would make preservation difficult. It would thus be the job of the Martian palaeontologist to search not perhaps directly for the fossils of organisms themselves, but instead for evidence that they may once have existed.

Searching for evidence of life, rather than for fossils themselves, is quite a common procedure for palaeontologists who study rocks that are older than the evolution of animals with hard parts on Earth. Wherever life has existed it will leave a mark of its presence, whether it is in the form of a physical mark, such as a footprint, or a chemical one, such as a carbonaceous smudge within a rock. In the ancient rocks of Earth we now have a number of means of detecting whether life was once present in a rock or not.

Among the easiest fossil structures to spot in the Earth's ancient sedimentary rocks are the aforementioned stromatolites. These are large structures that are made of layers of compacted sediment so that, in cross-section, they resemble a multi-layered dome. Their characteristic shape and large size are not caused by one organism, but by layers of single-celled algae that trap particles between them into layers that build upward to form their unmistakable rounded profile. Thus, when the stromatolite gets buried by sediment, although the algae may break down into nothing, the layered structure produced by them remains and is identifiable as a sign that life that was once present in that rock. Stromatolites, some of which can be found living in shallow seas today, have been clearly identified in rocks that are over 3,500 million years old on Earth.[47] Although we may not find actual stromatolites on Mars, we may find similar structures within the sedimentary rocks that indicate that some form of life was present. These may be layers of bound-together sediments, but they could equally well be pillars or balls of compacted sediment formed by organisms cementing grains together between them, or they could be areas of less well compacted sediment which has fallen into a space in the rock left by an organism that has rotted away. Whatever their form, such biologically produced sedimentary structures would be readily visible and identifiable to anybody exploring on the Martian surface, and perhaps even to a remotely operated vehicle fitted with a high-resolution camera. Other means of finding life would be trickier and would probably require actual Martian rocks to be physically examined by palaeontologists back on Earth.

There are two other possible means of detecting past life on Mars without the actual need for body fossils, both of which are dependent upon the body chemistry of the organisms concerned. Both these means are relevant to the story of the ALH84001 meteorite and, although mentioned below, are discussed more fully in Chapter 11.

On Earth, there are many kinds of bacteria that precipitate mineral grains as part of their body structure. The largest group of these organisms is known as the iron and manganese precipitating bacteria, which routinely produce the minerals pyrite and manganese within their bodies in the form of a sheath or simply as part of their cell structure. These minerals are much more resistant to erosion than body tissue and can survive the death of the organism to become part of the sediment. Such minerals, known as magnetofossils, can be identified as being biologically produced by their shape and chemistry and are thus a good means of identifying past biological activity. The occurrence of these magnetofossils is common in the fossil record, and some authors claim to have identified them in the Earth's oldest sediments at 3,850 million years in age.[188] Should the same adaptations have occurred on Mars, then such biologically produced minerals could be readily identified in the sediments there and may have already been done so in ALH84001 (see Chapter 11).

Alternatively, even if a primitive Martian organism could not precipitate minerals within itself, it would, probably, chemically alter its immediate environment so that it would leave a detectable chemical footprint behind it in the sediment. It could do this on a large scale, as is thought to have happened on Earth, by exhaling oxygen into a world almost totally dominated by carbon dioxide gas.

In a carbon dioxide-rich atmosphere, as existed on both early Earth and Mars, the presence of free oxygen would cause dissolved minerals, such as iron, to combine with the oxygen to form insoluble iron oxide which would gather on the beds of water bodies in large quantities. This is thought to have occurred on Earth approximately 2,000 million years ago, leading to the formation of massive, biologically produced iron deposits known as banded iron formations (see Chapter 13). These stand out in geological history because they are big and because they are an oxidised deposit formed at a time when oxygen was still very rare in the atmosphere. The association of the aforementioned magnetofossils and actual body fossils with banded iron formations suggests that the oxygen that produced them probably came from bacteria and that the formations are therefore closely associated with biological activity. The identification of features such as banded iron formations, or similar mineral oxides, on Mars could be evidence of past biological activity there.

The other, more subtle, means by which past biological activity could be identified is by differences in the way that biological organisms preferentially use different isotopes of the same element. Again, this concept is discussed more fully in Chapter 11, but it is worth outlining quickly here. Essentially, in terrestrial biological systems there are two different isotopes (i.e. versions) of carbon, a lighter (^{12}C) and heavier (^{13}C) isotope, available for use in life processes. Organisms prefer to use the lighter carbon isotope, which means that any biological matter, or sediments containing biologically derived matter, will have an increased amount of lighter carbon over heavier carbon by approximately 2 per cent.[143, 157] Measurements taken on Earth show that this isotope difference can be detected in biologically associated sediments of all ages, regardless of the changes in environment or atmosphere composition that has occurred over the aeons, suggesting that this is a reliable means of detecting life. A similar isotope shift has been detected in the Earth's oldest sediments, from Isua, Greenland, and also in the ALH84001 meteorite, and has been used to suggest the presence of biological activity there. Should any suspected organic matter ever be found on Mars, then the use of isotope analysis could prove to be a means of identifying it as being of biological origin.

As well as looking at purely Earthbound means of hunting for signs of ancient life, we must consider some more speculative ways in which Martian life may differ from that found on Earth. It may be, for example, that we will find body fossils on Mars. Perhaps they will be made of more resistant material than Earth's bacteria. After all, later in the geological record we get spores and pollen that are made of the incredibly resistant material kerogen, which could survive the majority of the processes liable to have been undergone by sediments deposited on Mars. It has always been assumed that Martian organisms would need to have been highly adapted to their less than friendly surroundings, which possibly means that they would need a thicker outer membrane (i.e. skin) to protect them against the increased cold, aridity or ultraviolet light. Such a membrane could also be better at fossilising than its terrestrial counterpart, and we could therefore find ourselves facing entire sediments made of dead Martian cells (unlikely, but still possible)! It may also be that, under the Martian gravity, which is two-thirds less strong than our own, organisms that are bacteria-sized on Earth would be able to grow significantly larger and therefore provide an increased amount of organic matter within the sediments.

All or none of these things may have happened; indeed, life may not have existed on Mars at all. All the questions regarding life, both past and present, can only be satisfactorily answered once we have actual Martian rocks in our possession. The prospects for this looked bleak until, in the 1970s, it was realised that we have had Martian rocks with us for thousands of years, and in 1996 it was announced by NASA that evidence of fossil life may have been found inside one of them.

11. ALH84001: THE METEORITE WITH A DIFFERENCE

7 AUGUST 1996

After the success of the Viking and Pathfinder lander missions to the Martian surface, it was natural for NASA to want to be able to bring actual samples of Mars's soil and rock back to Earth for a complete analysis. If we ever do get samples back, one of the first things that will be searched for would be any signs of life, and NASA has even discussed how such samples should be handled to prevent Earth becoming contaminated by a Martian organism.[99] However, the earliest possible sample return mission is not planned until 2003 and may not take place until 2005. In the meantime we already have a handful of Martian rocks here on Earth, and on the morning of 7 August 1996 I awoke to hear the radio news say that 'NASA scientists have announced that they have found evidence of fossil life on Mars.'

The story behind NASA's announcement of fossil life on Mars is long, complex and by no means certain. It centres around the discovery of chemicals inside an ancient Martian meteorite that may have been produced by living organisms and on the photographing of microscopic features in the same meteorite that show a resemblance to fossil bacteria. However, before NASA's examination of the Martian meteorites, there was a long tradition of studying meteorites in the search for extraterrestrial life. Many of the problems encountered in these previous studies were to shape the approach methods and presentation of the later NASA research. In fact, the NASA announcement of 1996 would never quite escape the stigma attached to extraterrestrial life and meteorites, particularly after one incident in the 1960s.

THE SEARCH FOR LIFE IN METEORITES

Organic matter was first recognised in a meteorite by the Swedish chemist Berzelius, who chanced upon a small meteorite that had fallen near the

French village of Alais in 1834.[20] He recognised that carbon-rich residues within the meteorite, similar in texture to soot, were not evidence of extraterrestrial life but instead were the product of complex chemistry that had taken place in deep space. Another chemist, however, studied an almost identical meteorite from Kaba and concluded that the soot was 'undoubtedly of organic [i.e. biological] origin'.[246] In 1864 another meteorite, containing the same sooty material, landed close to Orgueil, a small town in France. From then on recognition of this organic material within meteorites became commonplace, and the group as a whole were named the carbonaceous chondrite meteorites.

Interest in the carbonaceous chondrites built up slowly through the early twentieth century, but it was not until after the Second World War that people turned to these enigmatic rocks in the search for extraterrestrial life. In 1953 the chemical analysis of a carbonaceous chondrite showed that the soot-like material that defined the meteorites was composed of hydrocarbons, sulphur, nitrogen, oxygen and ash. All these chemicals are associated with life on Earth and gave hope that fossil life may have once existed inside the meteorite. A greater chemical breakdown of the organic material in the Orgueil carbonaceous chondrite by a series of researchers[26, 31, 56] revealed the presence of many of the building blocks of life, especially amino acids, sugars, purines, cytosine and imidazoles. Although many of these compounds were later proved to be terrestrial contaminants, at the time their presence inside a meteorite sparked a search for extraterrestrial life within the carbonaceous chondrites. In 1961 two respected scientists, Bartholomew Nagy and George Claus, made a discovery that was to grab the attention of the world in the same manner as the ALH84001 meteorite would do thirty-five years later.

In the science journal *Nature*, Nagy and Claus announced that they had discovered well-preserved microfossils in the Orgueil and Ivuna carbonaceous chondrite meteorites. They described finding particles, which strongly resembled biological organisms, embedded inside the meteorites, which were described by Urey as being 'organised elements'. The authors fancied that the 'organised elements' had features in common with modern algae and that they could recognise living structures within them such as double cell walls, spines, tubular protrusions, furrows, pores, vacuoles and flagella. A series of tests were performed on the organised elements, including staining to detect DNA (which was negative), dissolving in hydrofluoric acid to remove non-organic material (positive), high-density separation to concentrate the elements for study (positive) and an examination for optical polarisation which is common in living organisms (negative). The conclusion from this series of tests led Nagy and Claus to stand by their contention that they had discovered extraterrestrial life in the two meteorites.

Surprisingly, the rather obvious possibility of Earthbound chemicals and life working their way into the meteorites and contaminating them was brushed over lightly by the researchers, who argued that, whilst it was conceivable that bacteria and other nannolife could live within the rocks, single-celled algae, which the organised elements resembled, could not live in the dry environment of a museum where the meteorites had been stored. Strangely, the possibility of airborne contamination was not mentioned at all.

Nagy went on to provide secondary evidence for extraterrestrial life by suggesting that hydrated silicate minerals (i.e. crystals with water bound up inside their lattice) in the meteorite could only have formed in the presence of free water in a moderate pressure and temperature environment—in other words, the same conditions as are to be found over the majority of the Earth's surface today. There are not many places in the solar system where these environmental conditions exist, and although today the origins of the majority of meteorites are known, in the early 1960s they were not and the subject was still open to debate. To validate his theory Urey needed a plausible origin for the carbonaceous chondrites and their organised elements, and in 1962 he offered three possibilities. Even for their day, Urey's theories were somewhat surprising and caused a stir amongst scientists in all disciplines.

His first suggestion was that the carbonaceous chondrites could have come from a planet that once existed between Mars and Jupiter, where the current asteroid belt lies. The notion of such a planet, based on old mathematical models of the solar system, was once popular at the turn of the century but had more or less been completely disproved in the 1950s when it was realised that the asteroids were more likely to have been formed at the same time as the planets and were not the remains of an exploded planet. Urey himself concluded that this was the least likely of the carbonaceous chondrites' possible origins. The next obvious locality was the planet Mars, where many still believed life could be found. This, too, was discounted by Urey, because he felt that Mars's surface would consist solely of layers and layers of accumulated sediments and that the meteorites, having crystallised from liquid magma, could not have originated there. The third, and favoured, of Urey's possibilities seems naive and even comic, but it was taken seriously by many researchers at the time, including many at the newly formed NASA.

Urey turned his attention to the Moon as the body from which the carbonaceous chondrites originated and concluded that its closeness to

Earth and known geology made it the ideal candidate. This, however, still left him with the problem of the nature of the meteorites themselves, whose igneous rock type meant that they had crystallised from molten rock and not from accumulated sediments. On Earth, as we have noted, fossils are almost exclusively associated with sedimentary rocks which have been laid down under oceanic, lacustrine or atmospheric conditions, preserving the fossilised bodies of dead organisms between the layers of sediment. Urey knew that there was little evidence for the existence of sedimentary rocks on the Moon and that it was likely never to have had even an atmosphere, let alone liquid oceans. So how could igneous rocks from a sterile moon contain evidence of life within them?

Urey's theory was a novel one. Instead of postulating the existence of indigenous life on the Moon's surface, he suggested instead that the microfossils were of Earth origin. He envisaged that if a giant asteroid were to strike Earth's oceans hard enough, a tidal wave could be thrown up high enough to reach the Moon, 450,000 kilometres away. Pools of this terrestrial water, containing millions of planktonic organisms, would then slowly evaporate or seep into the lunar surface, leaving the fossils behind inside cracks within the Moon's igneous rocks. Another asteroid impact, this time on the Moon, would dislodge these microfossil-filled rocks, allowing some of them to return back to Earth, where they would become known as the carbonaceous chondrites. In effect, he was saying that the organised elements in the carbonaceous chondrites were fossils from Earth's distant past returning home again.

This theory received a mixed reaction amongst the scientific community at the time. Those that believed in the lunar transfer theory predicted that the Apollo astronauts would bring back Moon rock samples packed with microfossils from Earth's past. Others, such as astronomer Sir Fred Hoyle, confidently predicted that the Moon would be nothing but 'an interesting slag heap'. The debate reached its peak in 1962 when, in addition to the Orgueil and Ivuna meteorites, further organised elements were found by another research scientist in the Mokoia meteorite.[27] NASA, always keen on the thought of extraterrestrial life, formerly adopted Urey's Moon transference theory and used it as a motivating force for obtaining financial backing for the Apollo Moon landings. Fortunately for NASA and others, the mystery of the organised elements was solved long before Neil Armstrong and company brought back lumps of rock from the Moon to prove Urey wrong once and for all.

In the years following Nagy's, Claus's and Urey's announcements, a number of laboratories carried out in-depth analyses on the carbonaceous chondrites and found generally different results from those of the original investigators. Amino acids were found to be absent, as were nucleic acids, whereas non-biological compounds, such as melanina and ammeline, were found to be present, all of which suggests that life had not been present in the meteorite.[83] Another group of scientists found a close correlation between the hydrocarbons in the meteorites and those found in modern petroleum deposits,[148] although this was later explained away non-biologically.

Finally, the organised elements, the keystone of the extraterrestrial argument, were identified to the satisfaction of most scientists. Firstly, the organised elements that were arranged in crude hexagonal and other symmetrical shapes were demonstrated to be badly corroded crystals of iron sulphide.[11, 234] Then the cruellest blow of all came from an extremely thorough paper by Frank Fitch and Edward Anders,[66] who, amongst other things, correctly identified the most spectacular of the organised elements, the algae-like blobs, as degraded pollen from the ragweed plant. Ironically, this discovery was later reprinted in the same volume as a paper written by Nagy and others who were still promoting the extraterrestrial life theory for the carbonaceous chondrites. By the early 1970s thoughts of extraterrestrial life in meteorites were a distant memory as scientists concentrated instead on the data flooding in from the Mariner, Viking and Pioneer probes (see Chapters 6 and 7). Talk was beginning to generate about manned missions to Mars and the possibility of rock samples being returned,[167] but, as with most things that the human race attempts, Nature intervened.

THE SNC METEORITES

At the same time as the carbonaceous chondrite problem was being discussed and analysed, a separate group of meteorites had grabbed the attention of geologists. In 1968 it was noticed that three meteorites were made of the same rare igneous rock types and had all solidified from their liquid magmas at around the same time in geological history. These three meteorites had landed in the localities of Shergotty, Nakhla and Chassigny and as a result they where called the SNC meteorites—a name that has since been used to describe all meteorites belonging to that group.

Although all known meteorites are igneous in origin, it was noticed that the SNC meteorites were made of large, coarse crystals, indicating that they had cooled slowly inside a large, geologically active body, such as a planet, allowing the crystals time fully to develop. Most other meteorites are made of extremely fine-grained crystals, indicating that they cooled very rapidly, probably in deep space where the lower temperature causes molten rocks to freeze instantly. As well as being the same rock type, the

SNC meteorites also had a very similar chemical composition, particularly in a group of compounds known as the rare earth elements (REE), which effectively act as a chemical fingerprint for a rock. The three known SNC meteorites had very similar REE chemical fingerprints, which suggested that not only were they related to one another, but also that they were likely to have come from the crust of a planet rather than deep space.[61] The hunt began to find the host planet from which the SNC meteorites originated.

Isotopic examinations revealed that the SNC meteorites had all solidified from molten lava between 1,300 and 180 million years ago.[145] This age, although old compared to a human lifespan, is actually very late in the history of our solar system, which is approximately 5,000 million years old, and indicates that the host planet must have been large and geologically active enough to allow molten lava still to exist within it. A process of elimination began amongst the planets in the solar system. This was relatively easy as there are actually only a few which were still geologically active up until 180 million years ago. In fact, the only viable candidates are Mars and Earth. Both Jupiter and Saturn were ruled out as their size makes it impossible for small fragments to escape their large gravitational field and so no meteorite can leave their orbits. Venus was also excluded as its positioning in relation to the Earth makes it impossible for rock fragments to travel between the two planets. The SNC rocks are, geologically, completely incompatible with those on Earth, which left Mars as the only viable candidate.

Proving this was difficult until the results of the Viking landers' atmospheric analysis came through in 1976. Subsequent to this another member of the SNC clan, named EETA79001, was discovered in the Antarctic in 1979. This rock was found to have minute bubbles of its host planet's atmosphere trapped inside areas of it that had melted as a result of the impact that forced it off the planet in the first place. A study of the composition of this gas revealed that levels of isotopes of the gas argon were almost identical to those measured from the Martian atmosphere by the Viking landers.[22] This seemed too good to be true, but further analysis revealed identical ratios of other molecules and gases, particularly the oxygen isotope ratios, which are a key indicator of planetary origin. The issue was finally settled (as was Urey's earlier lunar transfer theory) in the early 1980s when the first true lunar meteorites were identified by comparison with the Apollo samples, showing that it was possible to identify the origin of a meteorite. By 1984 it was universally accepted that Mars was the host planet of the SNC meteorites, and by late 1993 eleven specimens were known. The existence of these Martian rocks was greeted with great jubilation amongst planetary scientists, who had obtained much needed Martian rock samples without the need for lengthy and expensive space travel. In early 1994 a geologist, working on a completely separate group of meteorites, chanced across a highly unusual example known as ALH84001. This was to be the start of a chain of events that was to end in the NASA announcement of fossil life on Mars.

THE DISCOVERY OF ALH84001

ALH84001 itself was discovered lying on the desolate and remote Allan Hills ice sheet during the 1984–85 meteorite field collecting expedition in Antarctica. Its discovery was unremarked upon at the time, it being a small rock 1.9 kilograms in weight and somewhat resembling an over-baked potato in shape and colour. An initial mineralogical examination of the rock identified it as a diogenite, a relatively common type found in meteorites originating from the asteroid belt. As such it was filed away with the hundreds of other known diogenites awaiting their turn for examination.

ALH84001 was next removed from storage in the early 1990s when a thin section of the rock was mounted on a slide as part of a routine examination of meteorite mineralogy under the microscope. A thin section taken from a rock will allow a very accurate identification of its minerals and, in this case, should have alerted geologists to its uniqueness. However, the slide taken from ALH84001 was mislabelled as coming from the diogenite EETA79002. The meteorite remained misidentified and was returned to storage again.[149, 193]

Shortly after this mistake another meteorite researcher, David Mittlefehldt, ordered a thin section from ALH84001 as part of a wide survey he was doing of diogenite meteorites. As soon as the slide from ALH84001 was placed underneath the his microscope he knew a big error had been made in its original identification. Instead of a fine-grained diogenite from the asteroid belt, Mittlefehldt was faced with a coarse-grained rock type known as a cataclastic orthopyroxenite, which indicated an entirely different origin for the meteorite. The large size and limited number of minerals in the rock indicated that it had cooled slowly at a high temperature. Following the logic that identified the host planet of the SNC meteorites in the first place, it became quickly apparent that this meteorite came from inside a planet and that this was most likely to be Mars. The same Martian REE fingerprint was discovered in ALH84001, and the exclusive SNC meteorite club gained its twelfth member. However, although it was of Martian origin, ALH84001 did not compare well with any of the other known SNC meteorites and so it became the only member of a sub-group within the SNC clan. Further work on its oxygen isotope ratio finally confirmed

ALH84001's Martian origin,[105] and in February 1994, almost exactly ten years after its discovery, NASA announced that a new and unusual Martian meteorite was available for scientists to study.

THE SEARCH FOR FOSSIL LIFE

The recognition of ALH84001 for what it was led to a flurry of excitement amongst the scientific community, many of whom immediately put forward proposals to NASA for studying the rock. Interest was such that the 1994 annual meeting of the Meteoritical Society was devoted entirely to ALH84001, its origins and implications. In the year after this results started to be published about the meteorite. It was at this point that a NASA scientist, David McKay, entered the story. His work on the meteorite went unobserved at first, but was later to put his name on the front page of every newspaper in the world.

In 1994 McKay was nearing the end of his time with NASA, having been recruited in the 1960s to teach geology to astronauts. He had been retained as a geologist and had since participated in a number of major NASA projects, including the Viking mission. He main interest was in the mineralogy of meteorites, and he had kept a close eye on the unfolding story of ALH84001 and its relationship to the SNC meteorites. Two early pieces of published research on the meteorite were to cause him to suspect that life could have once existed within it, and then persuade him to search for it within the rock itself.

The first piece of research to attract his attention was the estimated age of the meteorite. As was previously stated, the other eleven known SNC meteorites had been radiometrically dated as being between 1,300 and 180 million years old;[146] ALH84001 was placed at 4,560 million years,[77, 105] making it one of the oldest known rocks in the solar system. Although the planets are thought to have begun solidifying from the swirling dust clouds that formed them 4,600 million years ago, some, however, only solidified on the outside and remained hot and molten on the inside for millions of years. Earth and Venus still have molten cores, and Mars certainly did up until approximately 180 million years ago and is still partially molten today. On Earth, the constant movement of the crust, which is broken into small interlocking plates like a jigsaw, on top of the molten centres leads to old rocks being pulled under the surface and melted down. As a result there are very few rocks left on Earth that are over 1,000 million years old. The same process did not occur on Mars, meaning that large sections of its surface date back to the planet's early geological history—including ALH84001. It was exciting enough to most scientists for there to be a planetary rock from near the beginning of the solar system, but McKay realised that the rock would also have been around when Mars had a wetter, warmer climate possibly capable of supporting life (see Chapter 10). The possibility of fossil life existing within the meteorite entered his head but was, at that stage, treated as mere fantasy. The next piece of research to be published made him pay more attention to the idea.

In his original 1994 assessment of ALH84001, David Mittlefehldt had noticed that there was a secondary phase of more minor mineral growth within the meteorite, i.e., at some point after the rock had cooled and solidified more minerals had grown on top of the older ones. He commented that these growths were made of the minerals carbonate and pyrite and had taken place inside cracks and pore spaces within the meteorite. He also noted that 'some of the carbonates were clearly formed prior to arrival on Earth as the fine compositional zoning in them is offset along fine-scale fractures as a result of a shock'. He was essentially saying that the secondary minerals must have formed on Mars, rather than Earth, because they showed evidence of having been fractured by a meteorite impact or earthquake whilst still in the Martian crust. Carbonate and pyrite are common minerals that have been found in other meteorites (including the dreaded carbonaceous chondrites) but can also be associated with the by-products of life. However, Mittlefehldt effectively ruled out any biological origin for the minerals by calculating their formation temperature as being 700°C—far too high for the survival of life.

Subsequent to this estimate, a team of British meteoritic scientists, who were already studying other SNC meteorites, requested and received some precious fragments of ALH84001. The team, based at the Open University and Natural History Museum, had previously studied another famous SNC meteorite, EETA79001, and had found trace amounts of carbonate material which, when further analysed, showed signs of having organic molecules of unknown origin within them.[248] Although it was feasible that these molecules could have been biological in origin, the British team did not say so but instead ended their research by stating that 'the implications for studies of Mars are obvious'.[77]

A similar analysis was carried out on their ALH84001 samples, in which a number of crucial discoveries were made. The first of these was the sheer volume of carbonate material within the meteorite. In other SNCs the investigators had found levels of approximately 0.02 per cent (by volume) whilst in ALH84001 it was 1 per cent, much higher than had been previously found. In fact, the carbonates were so abundant that they formed into rounded, rosette shapes, later dubbed 'globules' by McKay, up to 0.2 millimetre across and thus visible to the naked eye. An oxygen isotope analysis was run on the carbonate globules to help gain an understanding of the conditions under which they formed. However, instead of getting Mittlefehldt's estimated formation temperature of 700°C, the oxygen isotope ratios suggested that it was between 0 and 80°C instead and that the globules had been formed by water from Mars's surface moving through the cracks and pores inside the rock.

Three of the main criteria set down by scientists for the existence of life were thus now fulfilled by ALH84001: first, the presence of the element carbon, the building block for terrestrial life, which was inside the carbonate minerals; secondly, the presence of liquid water, again essential for life on Earth; and thirdly, an equable temperature at which life can thrive. These conclusions were reinforced by another study of the globules which arrived at a similar conclusion: 'The oxygen isotope compositions of the carbonates indicate that they precipitated from a low-temperature fluid in the Martian crust. Combined with textural and geochemical considerations, the isotope data suggest that carbonate deposition took place in an open-system environment in which the ambient temperature fluctuated.'

Unbeknown to him, whilst McKay pondered the question of life in ALH84001, two other scientists had been thinking along the same lines. Chris Romanek, a geochemist from the University of Georgia, and Everett Gibson, a NASA meteorite specialist, had spent a considerable amount of time using a high-powered electronic microscope, called a scanning electron microscope (SEM), to study the surface of the carbonate globules and had found some unusual results that were hard to interpret. It was in the summer of 1994 that these two scientists approached McKay, who is an expert in SEM technology, about the possibility of collaborating in the search for life in ALH84001. The meeting resulted in a decision to form a team of specialists to search for different strands of evidence for fossil life having existed in the meteorite. It was also agreed that absolute secrecy was needed to prevent embarrassment should the results be negative—and, more importantly, to prevent anybody else from stealing their idea.

During the summer and autumn of that year another three people were requested to join the team. They were Kathie Thomas-Keprta, a tunnelling electron microscope (TEM) specialist from the Johnson Space Center and a colleague of McKay's; Richard Zare, developer of an advanced chemical analyser from Stanford University; and the latter's co-worker Simon Clemett. Work progressed slowly but steadily over the next 18 months, with each scientist working within his or her own specialised area of expertise to look for possible traces of past biological activity in and around the carbonate globules. The full evidence was assembled together in the early spring of 1996, written into a report by McKay, Gibson and others, and submitted to the respected American scientific journal *Science* on 5

April that year. After modifications and an intensive period of review, *Science* accepted the paper on 16 July and scheduled its publication date as 16 August. In the intervening time, *Science*'s editors decided that the sensational nature of the NASA team's findings meant that they should be kept secret until the publication date itself. A total embargo was therefore placed on any news of the research.

Before discussing the nature of the NASA team's findings, it should be noted that whilst they were fossil hunting, other scientists had managed to piece together a more or less complete history of ALH84001. An understanding of this rock's history was to be vital in the arguments that followed the *Science* publication, and is worth examining a little closer.

THE HISTORY OF ALH84001

Further work had confirmed the initial age of the rock as being 4,560 million years but that there was a so-called 'shock event'—i.e. a seismic disruption caused by an earthquake or meteorite impact—at around 4,000 million years.[12] It was this shock event that had caused the rock to crack and fracture considerably. Later, at around 3,600 million years (one source estimates this age to be 1,400 million years[238]), warm water from the surface is thought to have penetrated into these cracks and fractures and allowed the growth of the carbonate globules on the rock surface associated with them.[190]

One final Martian event was detectable in ALH84001. This was the shock event caused by the meteorite that dislodged the rock from the planet's surface and sent it through space towards Earth. This event, radiometrically dated as occurring 16 million years ago, also fractured the carbonate globules, making it certain that they were formed whilst ALH84001 was part of the Martian crust.[12] This was further confirmed when the quantity of interstellar radiation that the meteorite had adsorbed whilst travelling in space was measured—which also indicated that it had been in space for 16 million years. The final phase in the life of ALH84001 came when, 13,000 years ago, it wandered into Earth's orbit, was captured by its gravity and survived the heat of atmospheric entry to come to rest on the desolate ice fields of Antarctica. Thus, after such a complex life history, ALH84001 and the McKay team awaited the publication of one of the most controversial scientific findings of the century.

THE NASA ANNOUNCEMENT

Despite the secrecy of the researchers and the embargo by *Science*, rumours about the results began to circulate in early August, nearly two weeks before the publication date. Stories reached *Science* that a rival journal, *Space*

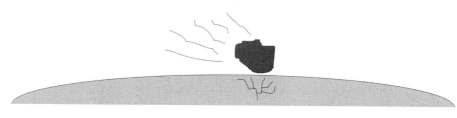

1. 4,600 million years ago: the Martian crust solidifies

2. 4,000 million years ago: a meteorite impact cracks the crust

3. 3,600 million years ago: life grows in the water-rich crust

4. 16 million years ago: another meteorite impact throws ALH84001 into space

Fig. 12. The geological history of the meteorite ALH84001.

News, was about to steal its thunder and make an announcement in its summer edition. To counter this *Science* immediately lifted the embargo on the paper and instructed NASA to hold a televised press conference, which had been originally scheduled for 16 August, a week early, on 7 August 1996.

Leading the conference was the Head of NASA, Daniel Goldin, backed up by the key members of the research team, including McKay and Gibson. Goldin began the proceedings by announcing, 'It's an unbelievable day. It took my breath away, but the scientists are not here to say that they've found ultimate proof and we must investigate, evaluate and validate this discovery.' After distributing advance copies of the paper to the press, McKay cautiously stated, 'We are not claiming that we have found life on Mars and we're not claiming we've found the smoking gun, the absolute proof, of past life on Mars. We're just saying we have found a lot of pointers in that direction. We welcome other people looking at [the evidence] and making other interpretations.' Thoughts of the 1960s carbonaceous chondrite incident were clearly not far from the minds of the scientific panel—nor, no doubt, from those of their critics.

Events moved fast. The announcement made the headlines on every news report and newspaper around the world. American President Bill Clinton said, 'Today rock 84001 speaks to us across all those billions of years and millions of miles. It speaks of the possibility of life. If this discovery is confirmed, it will surely be one of the most stunning insights into our universe that science has ever uncovered.'

After the world leaders, other people added their own voices and thoughts to the matter, all of which, in the early hours after the announcement, were based on gut reaction rather than informed opinion. By the end of that day every disc jockey had played David Bowie's song 'Life on Mars' (which, ironically, is actually about a visit to the cinema and not extraterrestrial life). Phrases like 'Little Green Men' were also flogged to death by all and sundry, including Goldin, who said at the conference, 'We are not talking about little green men . . . There is no evidence or suggestion that any higher life form ever existed on Mars.'

Despite the media circus that accompanied the NASA press conference, little more could be added to the story beyond the initial research findings and interest soon faded in the public eye, leaving scientists to fight amongst themselves over the McKay team's results. As these were only released on 7 August and not published widely until 16 August, there was a scramble by scientists (myself included) to read for themselves the evidence for fossil life on Mars. The paper was modestly entitled 'Search for Life on Mars: Possible Relic Biogenic Activity in Martian Meteorite ALH84001' and was deliberately speculative in its tone, the authors not wishing at any point to state that they had definitely found fossil life. The language used in the paper was even more guarded and cautious than normal for a scientific publication, making it clear at every stage that an alternative interpretation could exist for every conclusion. The presentation of the information was cold and clinical, almost as if the scientists were providing a recipe for a sponge cake rather than the possibility of life on other worlds. The paper focuses on five main strands of evidence, all based around the enigmatic carbonate globules, each of which are suggestive of Martian biological activity having occurred in the meteorite. The team concluded that, when taken together, the different strands of evidence all point directly to life having thrived on Mars over 3,600 million years ago.

THE EVIDENCE FOR PAST LIFE

As mentioned earlier, McKay and the other scientists concentrated almost exclusively on the nature of the carbonate globules in their search for life. These were only found within the fractures and pore spaces of the meteorite, indicating that they were formed after the initial crystallisation of the host rock and thus could be the relics of organisms that had grown within the meteorite. The first and most logical step was to analyse the nature of these globules chemically to see if they contained any hints about their origins. The earlier work by Romanek showed that the [13]Carbon isotope ratios were indicative of past life; now McKay and his team were searching for more complex organic molecules called polycyclic aromatic hydrocarbons (PAHs).

Although the term 'organic' is used to describe PAHs, this does not necessarily denote that they were produced by biological processes such as respiration, photosynthesis, organism growth or decomposition. In chemistry the term 'organic' actually refers to molecules that have the element carbon in their structure, and, as such, organic compounds can be produced non-biologically and therefore be found in sterile environments. PAHs are a classic example of this, and their presence has been documented everywhere from deep space to the ocean depths. The reason the McKay team were searching for these ubiquitous chemicals was because on Earth they are closely associated with the decomposition of living matter and as such are found associated with fossils. The discovery of PAHs in ALH84001 might well indicate that life had at some point been present within the rock but that it had now decomposed. Unfortunately, on Earth PAHs are also produced by the burning of fossil fuels and from decomposed organisms and may be carried around the globe in the atmosphere, making the possibility of PAH contamination within ALH84001 very real. With this in mind, it was instructed that, if PAHs were to be found inside the meteorite, all possibilities of contamination should be ruled out.

This task was left to Richard Zare and Simon Clemett of Stanford University, who used a recently developed, state-of-the-art microprobe two-step laser mass spectrometer to lift, separate and analyse individual molecules from the rock surface. The results showed that PAHs were common within the meteorite and that they were abundant within the carbonate globules themselves. A more in-depth analysis revealed that the PAHs in ALH84001 were present in two main chemical groupings of middle and high weights, the lighter group of PAHs being noticeably absent. This is significant, as the lighter PAHs, such as naphthalene, are characteristically found in space and are associated with meteorites coming from the asteroid belt. Their absence suggested that ALH84001's PAHs did not enter the meteorite whilst it was travelling through space. Instead, when they were compared with PAHs found on Earth, the simple chain-like compounds in ALH84001 strongly resembled those that were produced by the natural decomposition of biological cyclic compounds, i.e. living tissue. This led the team to conclude that they 'would expect that diagenesis of micro-organisms on ALH84001 could produce what we observed.' In other words, the PAHs in the meteorite were derived from the breakdown of living organisms.

The first question that this raised was the same one that, 30 years previously, had been brushed over so lightly, and with disastrous consequences, by the team examining the Orgueil and Ivuna meteorites. The PAHs may well be the remnants of tissue decay, but what if that decay was from terrestrial organisms that had grown inside the meteorite during its 13,000-year stay on the Antarctic ice? There was also the possibility that airborne terrestrial PAHs may have worked their way inside the meteorite. Consequently a number of tests were performed by Zare and Clemett to try to determine the origin of the PAHs.

The first test measured the levels of PAHs in the first 1.2 millimetres of the meteorite's outer crust. This revealed very low levels on the outside and in the first 0.4 millimetres of the crust, followed by a rapid increase deeper in the crust until they levelled off at 1.2 millimetres. Higher levels of PAHs on the inside of the meteorite than on the outside is the reverse of what would be expected if contamination had taken place on Earth: in such a case the molecules would be expected to be more abundant on the outside than on the inside as the contaminants worked their way in from the crust to the interior. The low level of PAHs in the actual crust itself suggests that those that were present on the outside of the meteorite had been vaporised during the atmospheric re-entry of the rock, again suggesting that the PAHs were in place inside the meteorite before it landed on Earth. Further tests were carried out on other Antarctic meteorites, some of them much more corroded than ALH84001; none of them contained similar levels or types of PAH. An analysis of local Antarctic environments and ice from Arctic cores also showed different levels and types of these omnipresent chemicals from those measured in ALH84001. All the evidence pointed towards both a biological and extraterrestrial, probably Martian, origin for the PAHs in ALH84001. The researchers had been extremely cautious and, with regard to the 1960s fiasco, McKay said, 'No one wants a repeat of that episode.'

The next task was to examine the enigmatic carbonate globules that drew the attention of the exobiologists to ALH84001 in the first place. Previous analyses had established that these features were circular or ovoid patches of zoned carbonate minerals that were restricted to the fractures and pore spaces of the meteorite. They ranged in size from 0.01 to 0.2 millimetre in diameter and were orange in colour with a distinctive black and white outer rim. It was also known that the carbonates had been formed in an aqueous environment between 0 and 80°C in temperature. One of the first observations made by McKay *et al* was that the globules were thin in nature so that they spread across the rock surface like a pancake. A suggestion was made that perhaps each globule had been forced to grow sideways, rather than outwards, in the restricted space within the meteorite's fracture surfaces. This shape was very suggestive of the way terrestrial organisms are forced to grow in confined spaces, although it should be noted that many minerals are capable of doing this as well.

Using the latest transmission electron microscope (TEM) and back-scatter electron (BSE) technology, both of which are capable of finely differentiating the chemical composition of substances such as minerals, some of the larger globules were examined in detail. Each globule was found to consist of a calcium (Ca)-rich core surrounded by a number of thin bands. These were alternatively rich in iron (Fe) and sulphur (S), and then rich in magnesium (Mg) with no iron. Further analysis of the mineral-rich bands showed that there was an outer and an inner band rich in the minerals magnetite (Fe_3O_4) and pyrrhotite (Fe_1-xS) embedded in a fine-grained carbonate matrix. The crystals from both these mineral types were described as being between 0.01 and 0.001 millimetre in length and of being cuboid or teardrop-shaped. When the centre of each globule was examined it was found that there were areas that were also rich in magnetite and an iron sulphide mineral suspected of being greigite (Fe_3S_4). More significantly, in the same area as these minerals the carbonate matrix showed signs of having been dissolved whilst the magnetite and greigite crystals remained

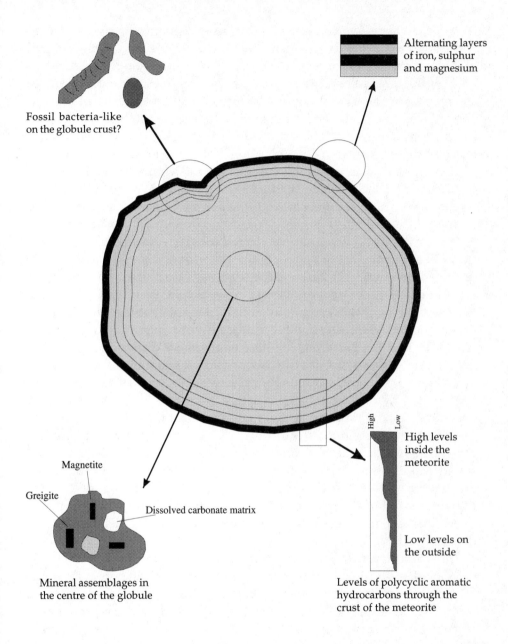

Fossil bacteria-like
on the globule crust?

Alternating layers
of iron, sulphur
and magnesium

Magnetite

Greigite

Dissolved carbonate matrix

Mineral assemblages in
the centre of the globule

High   Low

High levels
inside the
meteorite

Low levels on
the outside

Levels of polycyclic aromatic
hydrocarbons through the
crust of the meteorite

Fig. 13. The evidence for life inside a carbonate globule from ALH84001.

intact. In this lay another clue to the possible existence of ancient life within the meteorite.

On Earth, dissolved carbonate in the presence of intact crystals of magnetite and greigite can be produced by both biological and non-biological processes. The key to the argument in ALH84001 relies on the relative stability of the three minerals under acidic and alkaline conditions. Carbonate, magnetite and greigite can all be formed non-biologically in oxygen-poor alkaline conditions and remain stable together under such conditions. However, all three of the minerals are very unstable in acidic conditions and will dissolve very easily. This latter fact led to an interesting contradiction in the association of these three minerals within the globules.

Although all three minerals could have formed together, it would impossible, in natural circumstances, for one of the minerals to be dissolved without also affecting the others. Yet this is exactly what had happened with the carbonate, which showed heavy corrosion due to acidic dissolution whilst the magnetite and greigite, which should also have dissolved, remained untouched. In order for this strange situation to have occurred within the globules, a disequilibrium environment, involving a very high temperature of 700°C or more, would be needed. Yet the chemical evidence from the globules seemed to indicate a low formation temperature of 80°C or less. On Earth the only low temperature disequilibrium conditions that could produce the situation within the globules are associated with living organisms, particularly primitive ones such as bacteria.

Modern terrestrial bacteria have been observed to form both magnetite and iron sulphide minerals, including greigite, in the low-oxygen conditions that exist within their cell walls whilst at the same time dissolving other minerals, including carbonate ones, with other acidic processes. Could the disequilibrium conditions within the globules have been caused by Martian bacteria-like organisms? It was certain that the globules were Martian in origin, so contamination was virtually impossible, but it was quite a big leap to suggest they were biologically produced. Yet other evidence favoured this explanation as well.

The size, shape, chemistry and structure of the magnetite particles within the globules are directly comparable to terrestrial magnetofossils, which are the remains of magnetite crystals precipitated inside special bacteria called magnetosoames. On Earth, magnetofossils have been reported from ancient sedimentary rocks[43] and magnetosoames themselves are found in a huge range of environments from the sea bed to soil samples. The implication by McKay and his colleagues was that the globules, with their paradoxical mineral associations, PAHs, suggestive isotope ratios and biological looking shape, represented the broken-down remains of biological activity. The team's most startling, not to say controversial, findings were yet to come.

Since the 1960s there has been available to researchers a tool known as the scanning electron microscope (SEM). This device is essentially an extremely high-powered microscope that uses an electron beam to magnify details on samples many tens of thousands of times. In recent years SEM technology has improved considerably, allowing magnifications of hundreds of thousands times life size to be produced. Right from the start McKay had been using an SEM to look at sections of ALH84001, and he kept finding strange shapes that he could not fully explain with the equipment he had available to him. Fortunately, NASA had just invested in a new and exceptionally powerful SEM at the Johnson Space Center, and McKay duly took his samples to be examined under it. When he began to examine the surface of the carbonate globules he received quite a surprise.

Initially he examined the outer, iron-rich rim of the globules, which, at the highest magnification possible, he found to be made of thousands of microscopic, egg-shaped features less than 0.0001 millimetre in diameter. An examination of the central area of a globule revealed the same egg-like features there as well. These were a completely unexpected discovery and their existence was difficult to explain, especially as each 'egg' was far too small to be chemically analysed.

It was first suggested that they may have been the remains of dissolved carbonate, but dissolution in Earth specimens did not produce similar features. Laboratory or terrestrial contamination was ruled out after treating a number of other rocks, including meteorites, to exactly the same processing techniques as ALH84001 but with negative results. Again, a biological explanation was sought.

Even though they were aware of the controversy that would follow their thesis, McKay and the other team members suggested that the ovoid structures may be the fossilised remains of Martian life. Further searching with the SEM revealed not just ovoid structures, but more elongated tubular ones resembling flattened sausages. Most exciting of all was the discovery of tube-like structures that appeared to taper at either end and which apparently had a body that was divided into segments, like a worm. Pictures of these 'fossils' do indeed resemble modern bacteria in shape and form, and it is hard not to be impressed by them. In fact, when Everett Gibson took photographs of them home and put them on the kitchen table in front of his biologist wife, she is quoted as saying, 'What are these bacteria?'

This controversial claim was substantiated by making comparisons with modern nanobacteria (i.e. bacteria that are exceptionally small) living in hydrothermal springs, and it is indeed true that similar-shaped organisms do inhabit Earth and have done so since the dawn of life. Indeed, the remains of nannobacteria have been found fossilised inside rocks, proving that they can be preserved as fossils. The discovery of these features was the last strand in the team's conclusion that ALH84001 contained evidence of fossil life on Mars. The paper in *Science* ends by pointing out that non-biological explanations exist for everything the team had interpreted as being biological. However, the fact that so many different analysis techniques could produce results that essentially required a biological cause meant that it was at least possible, if not probable, that ALH84001 harbours evidence of past biological activity on the planet Mars. An invitation was issued by both NASA and all the members of McKay's research team for other scientists to replicate their findings and, if necessary, prove them wrong. This gauntlet was willingly taken up by a cynical scientific establishment eager to partake in the high-profile debate surrounding ALH84001.

12. ALH84001:
THE CRITICS AND THE FANATICS

THE DUST SETTLES

As soon as the dust had settled after the initial excitement of the NASA press conference, scientists from all fields began to take sides in the debate about the validity of fossil life in ALH84001. It was generally agreed that McKay's team had chosen the right approach and had used the correct methods in the course of their study. The cautious presentation of their results and the open invitation for comment was also welcomed by many as it gave the opportunity to criticise the results without entering into any personal battles. The first comments about the findings were generally 'soundbite' reactions from individual scientists who had been asked for a comment by newspapers or magazines. Many articles carried the famous quote by Carl Sagan that 'extraordinary claims require extraordinary evidence'. David McKay, and other members of his team, stated repeatedly that 'we want these results investigated and we are prepared to make samples of the rock available to credible researchers with sound experimental proposals'. In-depth research can, however, take months or years to complete and publish, and so the debate immediately following the press announcement was based almost exclusively on pre-existing research done on ALH84001 and the identification of ancient microfossils in general.

SCHOPF'S CRITERIA

The greatest critic of the possibility of fossils existing in ALH84001 is William Schopf, a University of California geologist who is a specialist in the search for, and identification of, some of the oldest fossils on Earth. As the majority of terrestrial rocks that are older than 1,000 million years have been severely altered by heat and pressure, the identification of fossils within them is difficult and problematic. Many claims of fossil finds in re-crystallised rocks of 3,000 million years or older rely on small, faint blobs or filament-like features that resemble organisms. Schopf is an expert at telling the real fossils from the false ones using physical and chemical means.

McKay, Gibson and Romanek invited Schopf to comment on their work months before the publication of their results. He responded that he could see nothing of any consequence in their research and even doubted that they would be able to get a publisher to take the paper. McKay openly refers to Schopf's negativity so early on in their research as 'a setback'. Nonetheless, Schopf agreed to keep his knowledge of the project to himself until any such publication occurred. This he did, but he was present, by invitation, at the NASA press conference, where he made his presence known to all. In the discussion phase of the conference, he made a number of extremely valuable comments about the meteorite's findings. 'We want to know if that organic matter [the globules] is demonstrably biological,' said Schopf, 'and secondly, with regard to the fossil-like objects, we'd like to know that they are assuredly fossils, not mineralic pseudofossils, or what we used to call foolers.'

Schopf's thirty years of experience in palaeontology and his famous debunking of a number of claims for the oldest fossils in the world made him well placed to ask these questions. He was also responsible for finding some of the world's oldest fossils, and as a result of years of research had developed seven criteria to which ancient fossils must conform to be considered genuine. He mentioned a couple of these at the conference itself, but, as the ALH84001 microfossils show many of the same features as those of ancient Earth, it is worth testing them using Schopf's criteria to see, had the claimed fossils been found on Earth, whether they would be considered representative of ancient life.

The first of Schopf's criteria states that the claimed fossils should be found in thin sections taken from a rock sample as well as in the rock itself. This is designed to exclude the possibility of laboratory contamination by demonstrating that the 'fossil' is embedded in the matrix of the rock and therefore part of it. In ALH84001 the carbonate globules are visible in thin sections although, as they occur in cracks, they are not embedded in the rock itself. Nonetheless, the globules show evidence of having been disturbed by the meteorite impact that blasted them into space, showing that they were not formed on Earth or in space. In this respect the globules pass Schopf's first test.

Secondly, the claimed fossils should occur in a sedimentary rock that has not been overly affected by heat or pressure. This is based on the assumption that sedimentary rocks represent the deposition of sediments in an aqueous or open-air environment and that any organisms living in these environments would need to become incorporated into the sediments in order to be preserved. Although ALH84001 is not a sedimentary rock, the fracture planes within it provide space where water and nutrients could have flowed and therefore allowed biological growth. Schopf acknowledges that intragranular spaces in igneous rocks may be suitable areas for biological activity. ALH84001 passes this test too.

Thirdly, the claimed fossil should be larger than the smallest free-living organism. This is based partly on the geological adage that 'the present is the key to the past' and partly on predetermined assumptions about the viability of microscopic organisms. The microfossils in ALH84001 fail this criterion quite badly. Although the diameter of the carbonate globules themselves is considerably larger than that of many living organisms, these have not been claimed to be individual organisms themselves but possibly the product of the breakdown of many individual organisms. However, the minute tube- and egg-like structures associated with the globules have been tentatively called 'microfossils', and these are considerably smaller than any known living Earth organisms. The longest of the ALH84001 microfossils is 0.0001 millimetre; the smallest known Earth organism is the bacterium *Coxiella* (a small, gram-negative pathogen) which may be as small as 0.002 by 0.004 millimetre.[229] This leaves a five-fold difference in size between *Coxiella* and the ALH84001 microfossils, meaning that they fail Schopf's third test.

Size is important to Schopf because, he argues, any organism with a length of less than 0.0025 millimetre would not be able to store, manufacture and catalyse the chemicals necessary to be viable. Organisms the size of those found in ALH84001 could only contain a maximum of 100 million atoms which, when compared to the hundreds of billions inside a spoonful of sugar, would limit the possibility for diverse chemical reactions. It must be admitted, though, that the size criterion is based on our knowledge of terrestrial, not Martian, biology.

The fourth criterion is that the fossil should be composed of kerogen, an insoluble biological material from which all modern microbes are composed. The presence of kerogen, which fossilises well, would thus set apart a potential fossil from a mineral that is shaped like a fossil. It is hard to know how strictly this should be applied to ALH84001 for, if the microfossils are genuine, then they may be composed of another organic material currently unknown to science. However, the principle of this criterion does apply and is central to the fossil argument. The globules are composed of minerals that may have been produced biologically or non-biologically. McKay's team, instead of finding kerogen, instead used a series of other tests to prove that there was an association with life in the meteorite. A whole suite of chemicals, including the PAHs, [13]Carbon, oxygen isotopes and mineral associations, within the globules is suggestive that they were once alive in a suitably low-temperature aqueous environment. It is uncertain whether ALH84001 passes this test, although, at the moment, it does have chemical evidence on its side.

Fifthly, the fossils should occur with others of similar size and shape. This is another Earth-centred criterion based on the theory that microbes will always grow in colonies. ALH84001 passes this with flying colours as both the globules and their microfossils occur in some abundance within the meteorite.

Schopf's sixth test is that the fossil should be hollow, which relates to his belief that hollowness is a key indicator of biological activity and that during the process of preservation the inside of the cell breaks down whilst the outside remains intact. The microfossils within the meteorite have so far not been observed to be hollow, although no broken specimens have been observed. Their small size makes it impossible deliberately to analyse or break apart an individual fossil to see what lurks inside. This issue was one of those raised by Schopf at NASA's press conference which prompted McKay to say that the finding of broken specimens would be a priority during further research.

Finally, the seventh characteristic is that the fossil should show some form of cellular elaboration. This is a highly debated topic, which states that fossils should have surface features such as spines, partitioning, binary fission or anything that differentiates them from smooth, round objects. The degree to how complex a 'fossil' has to be in order to be considered representative of a once-living organism is not clearly defined, and each case tends to be weighed up on its own merits. Some of the features photographed in ALH84001 do show characteristics which are reminiscent of cellular life. The most famous photograph of all, displaying an elongated and segmented tubular object resting on a sloping section of rock, looks uncannily like photographs of living bacteria. Other objects are more simple but have consistent shapes with strategic indentations that are also reminiscent of living bacteria. However, we again do not know exactly what it is we are looking for in relation to Martian biology, and trying to identify Earth-like features on potentially extraterrestrial organisms is not at all scientifically safe.

So, out of Schopf's seven criteria, the features in ALH84001 pass four of them, fail one and have two that are uncertain. In this respect they do quite well considering the unusual nature of the material being evaluated. The main problem with the microfossils is their small size. This means not only that they are prohibitively small compared to life on Earth but also that they cannot be separated from the host rock, split or chemically analysed, making it impossible for us to know what they are made of, how old they are or if they have any internal features. Even though they are life-like in shape and form, the possibility exists that they are no more than corroded minerals, dust, clay particles or even lumps in the ultra-thin layer of gold that has to be applied to all objects before they can be examined under an SEM. McKay and colleagues did everything in their power to minimise contamination but also readily admit that the microfossils are the weakest argument that they have presented in regard to ALH84001. Instead, the crux of the team's argument rests on the chemical analysis and mineral associations within the carbonate globules. Other scientists, too, concentrated their efforts on this area, the first results starting to appear in a number of small-circulation scientific journals during the opening months of 1997.

OTHERS JOIN THE DEBATE

Despite the wide range of results presented in the original *Science* publication from August 1997, the majority of scientific comment and research has concentrated on the temperature of the water in which the carbonate globules formed on Mars. This, in Earth terms, is a crucial issue for the survival of life: too hot, and the organisms will be boiled apart; too cold, and their metabolism will slow down and freeze. Most living organisms live within a tight 0–40°C band, although some can survive in temperatures as low as –40°C and others in temperatures as high as 110°C. These latter extremes are generally superimposed as being the thermal limits within which life can survive on other planets. The logic behind this is not entirely clear, as we cannot possibly know extraterrestrial biological requirements. Even so, in the case of the Martian meteorite, temperature has been made a vital issue and is therefore important in terms of its analysis.

It is already known that the carbonate globules were formed in an aqueous environment. However, if the water temperature was low (less than 80°C) and the water originated from the planet's surface, as the oxygen isotopes suggest, then the rock itself was probably not buried that deep within the crust. This makes it possible for nutrients to be brought down to any organisms that may thrive within the cracks and large pore spaces of the rock in the same manner as modern bacteria pulled from terrestrial boreholes many kilometres deep (see Chapter 9). It also means that the combination of dissolved carbonate and intact magnetite and greigite would be impossible to form without the aid of biological activity.

A much higher water temperature would imply that the hydrothermal fluids percolating through the material were derived from deeper in the mantle or possibly from the impact of an asteroid or similar body. A high temperature, in conjunction with a higher pressure, would also allow the carbonate, magnetite and greigite combination to form. The formation temperature of the carbonate globules is therefore crucial to the validity of the claim for past Martian life in ALH84001.

The initial diagnosis of the globules was made by David Mittlefehldt, who used microscope observations of the relationships between minerals to deduce that they would be unstable at lower temperatures and would have needed an environment of 700°C in order to form stably. However, that same year another group of scientists, some of whom would later become part of McKay's team, used oxygen and carbon isotopes to determine that the temperature formation was between 0 and 80°C and that the water involved had at some point adsorbed carbon dioxide gas from the Martian atmosphere, indicating that it had originated from the planet's surface.[190] Work done one year later confirmed the [13]Carbon isotope levels found previously and suggested that the globules were of extraterrestrial origin as opposed to being the result of terrestrial contamination.[107, 108]

The next publication on the subject of the globules came out marginally before the announcement of fossil life made by NASA in 1996. This research analysed the major element composition of the carbonates in an attempt to reconstruct the original formation conditions of the minerals.[81] The observed chemistry of the carbonate minerals showed that no water had been incorporated into their crystal lattices. This led to the suggestion that the globules probably formed by rapid cooling from high-temperature fluids produced in the aftermath of a meteorite impact; in other words, the carbonate globules were not biologically produced but the product of normal crystallisation. To overcome some of the other pro-fossil arguments used by McKay, it was suggested that a nearby meteorite impact could produce hot (650°C) fluids saturated in atmospheric carbon dioxide which would percolate into the crust. Here a reaction between the rock and the hot fluids would cause the carbonate globules to form, with a significant quantity of atmospheric oxygen and carbon isotopes within them. A similar conclusion was reached by another scientific team led by a University of Hawaii geologist, who proposed that the carbonate globules were formed as the result of a high-temperature shock from a meteorite impact.[213] To contradict these two theories, another study did find significant amounts of water trapped within the globules and studied it for its hydrogen isotope levels.[124] All further analyses were to be published after the McKay paper and were therefore very aware of the biological implications attached to the carbonate globules.

The first proper criticism of the fossil theory appeared in the form of a letter published in the journal *Geochimica and Cosmochimica Acta* a matter of weeks after the NASA press conference. The research used yet another set of isotopes, those of the element sulphur, to try to determine the temperature and mode of formation of the globules. The level of a sulphur isotope that was found within the greigite crystals suggested that either they had formed under low temperature, low oxygen and alkaline conditions or that water from the planet's surface, enriched in the sulphur isotope, had percolated into the crust prior to the carbonate globules' formation. Either way, the elevated levels of the sulphur isotope were deemed to be inconsistent with pyrite forming biologically.

In the meantime, one of the teams that had first alerted McKay and Gibson to the possibility of fossil life occurring within ALH84001 had done further work on ALH84001 and another of the SNC meteorites, EETA79001, the results of which were released in November 1996. The research team, based in Britain, worked specifically on both meteorites' oxygen and carbon isotopes, and the members were clearly keen to be associated with the work of the NASA scientists, even holding a press conference to release their results. At this conference a press release was handed around which started with the words 'Today . . . [we] offer the strongest support yet for the hypothesis that life once existed on the planet Mars.'

Their research on ALH84001 concentrated on the ratio between light and heavy carbon isotopes (^{12}C and ^{13}C), which, since biological activity selectively uses heavy carbon, leaving its lighter cousin behind, is a good indicator of biological activity. In one of their samples they found a dramatic depletion of heavy carbon levels, making the samples twice as light in heavy carbon as is found in most Earthbound organic matter. On Earth the only way for this to occur is through the production of methane by bacteria. The team also referred to earlier research done on EETA79001 which had found small concentrations of the contentious PAHs, used as evidence of life within ALH84001. It was therefore concluded that evidence for fossil life had been found in ALH84001 *and* EETA79001! As EETA79001 was only blasted from Mars 600,000 years ago, this meant that life on Mars could have been present up until quite recently.

An unpleasant debate followed the press conference during which the British scientists were subtly accused of publicity-seeking: 'British Scientists Seek Recognition of Role in "Life on Mars" debate', ran the headline in *Nature*. The scolding they received may, however, have been justified when the team, some months later, privately admitted that on further testing they had had problems with their figures. 'After October 1996, we are happy to conclude that some of our initial isotope fractionations resulted from an analytical artefact,' confessed Monica Grady, one of the team's members. The failure of this research meant that by Christmas 1996 there was a balanced amount of research in favour of and against the existence of fossil life in ALH84001.

In the following months other scientists declared or published their own carbon and oxygen isotope data, most of which broadly confirmed a lower temperature of formation than McSween's original estimate of 700°C. Interestingly, A. Jull[108] also found the same carbon depletion in the globules as that seen by the British scientists, and cited either the Martian atmosphere or biological activity as the cause. Hutchins and Jakosky[100] re-examined the oxygen isotope data and derived a formation temperature of 40–250°C which, in its lower half, overlaps the 0–80°C estimate. An almost identical study again found 'no evidence for high-temperature carbonate precipitation'[235] and concluded that the temperature would have to be lower than 300°C. One further piece of research, which did not rely on isotope data, confirmed the likely lower temperature formation of the carbonate globules by examining the magnetic alignment of minerals in the host rock of the meteorite.[117] Any exposure to even relatively moderate temperatures would have destroyed this alignment, but, as the alignment was intact, it was concluded that the meteorite had not been exposed to high temperatures. Things were looking better for the prospects of fossil life having existed inside ALH84001.

All the research so far had concentrated almost exclusively on the oxygen and carbon levels within the globules, yet this was only a fraction of the evidence presented by McKay et al in 1996. In particular, the issue of the PAHs and their origin seemed to have been completely ignored. Unfortunately for the pro-biological camp, a devastating new piece of research was unveiled in early 1997.

Three researchers[17] analysed another SNC meteorite, EETA79001, for PAHs and found them throughout their samples, suggesting that PAHs were not unique to ALH84001. Next they measured levels of PAHs in the polar ice and found that carbonate minerals within the ice were very effective scavengers of PAHs from the surrounding atmosphere. An examination of the types of PAH in the ice showed that they were similar to those in both ALH84001 and EETA79001. It was therefore suggested that the PAHs in both meteorites had been placed there from meltwater running through the centre of the rock and that the carbonate globules would be more effective at scavenging them than normal rock, thus increasing the levels of PAH found within them. As if this were not bad enough for McKay and his fellow scientists, the same research also produced trace amounts of L-amino acids in the meteorites. These were unquestionably of terrestrial origin and very strongly hinted that the inside of the meteorite had indeed been contaminated during its long stay on the Antarctic ice. The conclusion was that 'PAHs are not useful biomarkers in the search for extant life on Mars.' It should be noted that the NASA team had anticipated much of this research and performed similar tests on the distribution of PAHs in ALH84001, other meteorites and polar ice cores, all of which were negative. They did not, however, search for amino acids.

So, in the months and years after their announcement, how well have McKay's results stood up to scientific scrutiny?

THE CONCLUSION THUS FAR

The accompanying table shows the number of results published on the meteorite and their conclusions on the existence of fossil life as defined by NASA findings. This demonstrates the limited amount of work that has been done on the meteorite, almost all of it concentrating on the temperature at which the carbonate globules were formed. In terms of numbers, there is a general agreement that this temperature formation was much less than that initially suggested (700°C), although a general figure has not yet been settled. Other work on the light-versus-heavy carbon ratios does also suggest that some heavy carbon depletion has occurred and could be due to biological activity, although other explanations are also possible. The only absolutely negative results were those concerning the PAHs and

---

Summary of Research Performed on Globules within ALH84001 and their Implication for the Possibility of their Representing Fossil Life

Research team	Method used	Major results	Fossil life in ALH84001?
Mittlefehldt[149]	Mineralogy of the globules	Formed at 700°C	Not possible
McKay et al[144]	Carbon isotopes, PAHs, microscope examination	Low temperature formation, possible biological features	Possible
Scott et al[213]	Mineral chemistry of the globules	Formed at 650°C	Not possible
Leshin et al[124]	Sulphur isotopes	Low temperature formation	Possible
Harvey and McSween[81]	Mineral chemistry of the globules	Formed at 650°C	Not possible
Jull et al[108]	Carbon isotopes	Low temperature formation	Possible
Hutchins and Jakosky[100]	Oxygen isotopes	Formed between 40° and 250°C	Possible
Valley et al[235]	Oxygen isotopes	Less than 300°C	Uncertain
Krischvink et al[117]	Magnetic mineral alignment	Low temperature formation	Possible
Becker et al[17]	PAHs, L amino acids	Globules contaminated from polar ice	Not possible

amino acids within the meteorites and local environment, although McKay's original research does refute some of these findings.

There are still large areas of McKay's research that need to be re-examined (for example, the disequilibrium mineral associations), and the issue has still not been conclusively resolved one way or the other. The vested interests of some scientists and the sensitivity of the whole topic means that a satisfactory answer will not be reached for some time yet and may only be forthcoming when the first Martian rocks are returned from the planet itself. That, however, is a long way off, and for the moment the issue will probably only be settled by a majority of scientific consensus rather than any one piece of research.

FULL CIRCLE

This debate about the possibility of fossil life being preserved within meteorites began with the carbonaceous chondrites and in particular the now infamous 'organised elements' mistake of the 1960s which has overshadowed the ALH84001 results. It is interesting, therefore, that almost exactly a year after the NASA announcement about ALH84001, another planetary scientist has again claimed to find signs of life in another carbonaceous chondrite meteorite that fell near Murchison in 1969.

The Murchison meteorite is the best preserved and least contaminated of the carbonaceous chondrites, and the scientist concerned, Richard Hoover, claims to have found tiny spherical and spiky structures within the centre of it. He says that 'although most are probably . . . of non-biological origin, some populations have distributions, chemical and morphological characteristics suggestive of coccoid bacteria and cyanobacteria'. He also describes other features that look like slime mould similar to the globules in ALH84001. His interpretation of these structures is that they may be fossils, possibly ejected from Mars or Earth, that flew through space on chunks of rocks which may eventually have crashed into the Murchison meteorite, contaminating it before its descent to Earth.

As Urey found in the 1960s, explaining the presence of life in a deep-space meteorite is difficult, and reaction to Hoover has so far been sceptical. Another geologist, John Kerridge, has commented that 'all Hoover's got is shapes. Like many people, he seems to believe that minerals have crystal faces, and anything without nice crystal faces has to be biology.'[44]

13. THE ORIGINS OF LIFE ON MARS AND EARTH

In Chapter 10 the likelihood of life evolving at all under the past environmental conditions of Mars was discussed. Given the presence of water and other factors on Earth and Mars, it is reasonably certain that, whether it did not or not, suitable conditions existed whereby life could have evolved there.

Whilst the exact shape, form and function of any evolved life on Mars can only be guessed at for now, there is a large body of opinion that believes that life on Earth and Mars would be identical. Some have even suggested that life could have travelled between the two planets and that the possible discovery of fossils in ALH84001 proves a link between the two planets.

ORIGINS OF LIFE

The first question to ask is, what is the likelihood is of identical life evolving separately on two different planets at the same time? As discussed already, between approximately 4,500 and 3,800 million years ago the environments and geological history of both planets were broadly similar. It is therefore also likely that the same physical geographical processes—erosion, weather conditions etc—would also have operated on both planets and the same basic chemicals would also have been present. In these circumstances it is perhaps possible, or even probable, that life could evolve on both planets. But identical life?

Entire journals are devoted to the issue of how life first originated, and it is beyond the scope of this book to enter into the vast debate about how a group of complex chemical compounds gathered together inside an organic coating could have become a reproducing organism. Even so, I shall briefly discuss some of the theories that are currently in vogue amongst evolutionary biologists, although the reader will need to consult other textbooks for more detailed information. There are currently two main groups of theories concerning the origin of life. One states that life evolved on each planet separately, the other that it evolved elsewhere in the solar system or universe and was seeded on to the planets from space.

THE EVOLUTION OF LIFE ON EARTH

It has always been felt that life must have originated on Earth after the formation of its oceans, approximately 4,000 million years ago, but before the first signs of fossil life, definitely 3,500 million years but possibly 3,850 million years ago.[89] The end of the bombardment of the planets by meteorites, some of which could have vaporised entire oceans, occurred approximately 4,000–3,800 million years ago and, again, would probably have placed a restraint on the time from which life could have evolved.[135] Thus, depending on these factors, there is a short period of approximately 100–500 million years on Earth when life could have evolved. On Mars this would be slightly less as the surface water was beginning to disappear rapidly by 3,800 million years ago (see Chapter 10).

Before life could originate, a number of requirements would have to be in place, the most important of which are water, organic compounds, an energy source and a site in which to concentrate them. All of these would have been available on both Mars and the Earth. The water and a carbon dioxide atmosphere would have come from volcanic gases, the organic material from lightning strikes and carbonaceous material brought in by meteorites (such as the carbonaceous chondrites; see Chapter 11) and comets hitting the Earth and the energy from sunlight or the heat from volcanoes, hot springs or other geological processes. Once these basic requirements had been met, there must have been four physical processes before life could have originated: the organic compounds must have accumulated together, they must then have formed larger, more complex molecules which must then themselves have formed into a early protocellular structure, and finally a self-contained system must have originated that could have taken in energy and nutrients from the environment in order to reproduce itself.[55] There is much debate about how and where these processes operate.

A much-favoured theory is that the assembly of life took place in a shallow tidal-pool environment where the incoming tide would fill the pool with water rich in organic compounds such as amino acids and proteins. It is then speculated that when the tide went out again the heat of the Sun would have evaporated away much of the water, forcing the organic chemicals to concentrate together to form more complex organic molecules, possibly by polymerisation reactions involving amino acids and nucleotides.[55] Eventually these larger molecules would have become trapped inside an air bubble, thus forming the earliest self-contained cell.

In recent years this theory has gone out of vogue considerably, with the discovery by geneticists that our earliest ancestor was probably a heat-loving bacterium living around a hot spring. This has brought people to wonder about whether another recently discovered geological phenomenon, the deep-ocean black smoker vent, could have been a likely place for life to evolve. In theory black smokers are ideal as they are warm, belch out abundant nutrients and chemicals from the Earth's interior and would be surrounded by water at the bottom of an ocean. It has also been speculated that any primitive life in a black smoker could have survived the early meteorite bombardment, and that therefore life could have originated as long ago as 4,200 million years.[135] It is thought possible that hollow minerals of pyrite (iron sulphide) could have provided suitable 'reaction vessels' within which organic molecules could have accumulated to form an early cell, although no laboratory work has yet been done on this.[80] Other sites for the evolution of life include volcanic and meteorite crater lakes and comet impact points, although none of these are currently thought likely.

The means by which 'life' jumped from a group of associated organic molecules to a replicating organism is highly complex and will not be covered here (see Cairns-Smith[30] or Fortey[73] for a better coverage of this topic). Instead, it is the aforementioned environments and circumstances in which life could have arisen that is more relevant to the 'life on Mars' issue. As we currently understand it, on early Mars there was water, volcanism, organic chemicals, light and heat. There is thus no reason why life could not have evolved on Mars at the same time as it did on Earth. If life formed by the same mechanism as it did on Earth, which ended in the evolution of RNA and later DNA, then perhaps we should expect similar organisms to result on Mars. All current evidence indicates that the earliest life forms on Earth were single-celled bacteria that floated in a primitive ocean or rested on the sea floor. It is therefore likely that any life on Mars would be similar to this, and, indeed, if the NASA results are genuine, there would appear to be a similarity between the ALH84001 'fossils' and early life on Earth

SIMILARITIES BETWEEN ALH84001 AND EARLY EARTH

In the following section it will be assumed that the evidence for life in ALH84001, as presented by David McKay and others,[144] is genuine (something that has not been established for certain; see Chapter 12). Even if the ALH84001 evidence is found to be false, the following comparison to early life on Earth will serve as a useful guide to what we should expect to find on Mars in terms of fossil life if the mechanism for the development of life was the same on both planets.

The oldest known body fossils on Earth are 3,450 million years old, which is only marginally younger than the age given to the Martian globules. These fossils were found in a series of cherty rocks located near Port Headland in Western Australia and were studied by Schopf and Packer.[209] The fossils themselves are small, being described as 'sheath-enclosed spheroidal cells', about 0.01–0.003 millimetre in diameter, some of which form into colony-like structures. Many analogies were found to a modern bacteria group, called the cyanobacteria, which produce oxygen via photosynthesis, and it was concluded that their microfossils may have been early representatives of this group. Other fossils of similar age have also been related to bacteria of various forms, including a peculiar group of bacteria which form massive geological structures known as banded iron formations (BIFs). This latter group may have some significance to our story and will be discussed in detail later.

We know much about these early terrestrial microfossils through comparisons to their modern relatives. Although the ALH84001 microfossils may possibly have living representatives on Mars, we have not found them yet and so understanding what exactly the meteorite's fossils represent is made much more difficult. Even so, there are a number of obvious points that can be made about them, based on the limited information gained by McKay and others.[144]

First, the living organisms represented in ALH84001 must have been small and self-contained in nature. The maximum size of the carbonate globules is 0.25 millimetre, and, as these are supposed to have been the product of living organisms, any life they represent must have been this size or smaller. In point of fact, the microfossils photographed on the globules themselves are very small indeed, being only 0.0001 millimetre or less in length.

The globules themselves are separated from each other within the meteorite, indicating that the organisms were capable of functioning independently of each other and may have had some form of outer cell wall within which biological reactions could occur. The minute fossils on the globules are certainly self-contained and seem to have retained a basic cell shape after fossilisation. Other points to make are that the organisms must have grown *in situ* as the globules are shaped to fit in the narrow fracture and pore spaces in which they are found. We also know that they grew in temperate water, in a carbon dioxide-rich atmosphere, and that they survive without the need for direct sunlight, which was probably unavailable inside the Martian crust where ALH84001 originated. However, to understand exactly how the ALH84001 fossils functioned we need to look at the geochemical analysis of the globules and their mineral associations. Here we find quite uncanny similarities to what we know about life on Earth at the time.

The heavy-versus-light carbon isotope ratio in the globules is the same as that produced by modern bacteria and, more crucially, is within levels measured in rocks up to 3,850 million years old on Earth. This not only shows the possible importance of carbon as a basic building block for life on Mars, but also indicates that the Martian organisms may have lived by preferentially fractionating light carbon from the carbon dioxide-rich atmosphere of Mars. Earth rocks of the same age show precisely the same carbon isotope ratios, and at that time Earth had a carbon dioxide-rich atmosphere identical to that of Mars.

There are also the associations of the minerals magnetite and pyrite in the carbonate globules, an association that, McKay infers, has a biological origin. It is certainly true that these minerals are produced within modern bacteria and are known to have operated in the fossil record as far back as 3,500 million years and possibly to 3,850 million years ago, which again overlaps with the age of the Martian fossils.

The last, and most obvious, comparison between Earth bacteria and the ALH84001 microfossils is that the scanning electron microscope photographs show features that are remarkably bacteria-like in shape and appearance, albeit many times smaller than any known Earth bacteria. All in all, the similarities between the Martian microfossils and those that were around on Earth at about the same time are great. There is, however, one group of fossil bacteria to which they show a particular affinity.

Banded iron formations (BIFs), which have already been briefly mentioned, are one of Earth's great mineral treasures and provide the staple source of iron ore in many parts of the world. BIFs are geological structures which may be many hundreds of metres thick and thousands of kilometres in lateral extent. They are characterised by having alternating beds of siliceous-rich (e.g. chert) and iron-rich sediment. The iron minerals which dominate the BIFs are magnetite and iron sulphides (such as greigite), the same as those found in the globules. The majority of BIFs were laid down in a short time interval between 2,600 and 1,800 million years ago, but they are known to have existed 3,850 years ago, 250 million years before the ALH84001 globules. It is now generally agreed that BIFs are the result of bacterial action, formed possibly by the direct precipitation of minerals within the bacteria themselves or by the oxygen they produced causing dissolved iron in the surrounding water to precipitate out on to the sea bed. Indeed, it is thought that the oxygenation of the Earth's atmosphere, which occurred approximately 1,500 million years ago, led to the demise of the BIFs in the geological record.

A study of the bacterial fossils associated with the BIFs reveals similar-shaped organisms (although bigger) to those which have been observed inside the meteorite, with similar mineral and chemical associations. It is, of course, pure speculation to suggest a firm link between the BIFs and features in a Martian meteorite that have not yet been confirmed as fossils, but nonetheless the similarities are there. Some would argue, possibly correctly, that any similarities between the ALH84001 globules and life on Earth have arisen because the globules are due to contamination from the Antarctic ice, others that any similarities are purely coincidental and that the globules are of non-biological chemical origin. On the other side there are those who have predicted that any fossil, or even current, life on Mars would be identical to that on Earth due to a process called panspermia. Panspermia is the second group of theories that has been used to explain the origins of life on Earth, and one of its predictions is that any life in the solar system or even the universe may be similar.

PANSPERMIA AND THE ORIGIN OF LIFE

The underlying thesis of panspermia is that life was seeded on to the Earth and other planets from space. There are two main theories as to how this could have happened: the first is that life could have developed in outer space and then landed on Earth, the other that life evolved on another planet and then travelled through space to Earth.

Supporters of life originating in space and then landing on Earth believe that many of the chemicals necessary for life, or even primitive life itself, were seeded on to planets such as the Earth and Mars by passing comets. This theory was chiefly developed and pioneered by two British astronomers, Fred Hoyle and Chandra Wickramasinghe, during the 1970s and 1980s.

LIFE FROM COMETS?

Hoyle's interest in comets and the possibility of life was stimulated by a curious chapter in a book about the common cold that was published in 1965.[6] The book's author, Christopher Andrewes, noted that the spread of cold and flu viruses across the world was not uniform or predictable. Andrewes thought that if a new strain of a virus appeared in one region of a country then it should spread outwards from the centre of its first recorded occurrence like ripples moving outwards from a dropped stone in a pond. However, the pattern that was actually observed was that the same virus would appear simultaneously in different parts of the world and not at all in others. The rate of infection was also strange, with viruses moving like wildfire across some sections of a country and not spreading at all to others. To Andrewes, and also to Hoyle, this simultaneous appearance and uneven rate of spread seemed illogical for a person-to-person, droplet-spread infection.

Hoyle puzzled about this conundrum and thought that perhaps the viruses were being distributed on an intercontinental scale by airborne means. The problem is that for this to happen the viruses would need to travel within the strong jetstream air currents that circumnavigate the world at a height of 10,000 metres or more. It was difficult to explain how such viruses could have travelled from the surface of the Earth into the jetstream in sufficient numbers to allow them to reach other continents. Hoyle concluded that, perhaps, instead of entering the jetstream from Earth below, the viruses were entering it from above and then getting scattered by the winds across the countryside. As viruses are not known to survive in space on their own, and as they would not survive entry into the Earth's atmosphere without protection, Hoyle speculated that the viruses must be travelling inside comets.

Comets are thought to be remnants of the original clouds of ice and dust from which the solar system formed approximately 5,000 million years ago. They are often described as being 'dirty snowballs' and are known to be balls of water and carbon dioxide ice with small fragments of rock mixed within. They travel in elliptical orbits around the Sun which sling them from the outer reaches of the solar system to within a short distance (astronomically speaking) of the Sun itself. During its approach past the Sun the 'dirty snowball' becomes heated, causing it to develop a following tail of melted gases and dust grains. The effect of a comet's tail on the Earth is considerable. Not only are the tails visible from Earth, but they leave permanent trails of ice and dust particles behind them. When the path of Earth through space crosses the remains of a cometary tail, the particles of ice and dust enter the Earth's atmosphere, where they burn up to form shooting stars. Hoyle speculated that some of the larger particles do not burn up entirely and make it into the upper atmosphere, where they release organic matter and even viruses which make their way down to the surface via the perpetually operating weather systems.[94, 95, 96] But how did organic molecules, let alone whole viruses, become incorporated into the comets in the first place?

This was where the work of Wickramasinghe comes into play. He believes that at the solar system's birth, when it was essentially a large dust cloud surrounding the early Sun, the action of ultraviolet light on particles of carbon, iron, water and hydrogen would create complex organic molecules which would float freely in space. These chemicals are known to have existed in the beginning of the solar system, and organic molecules,

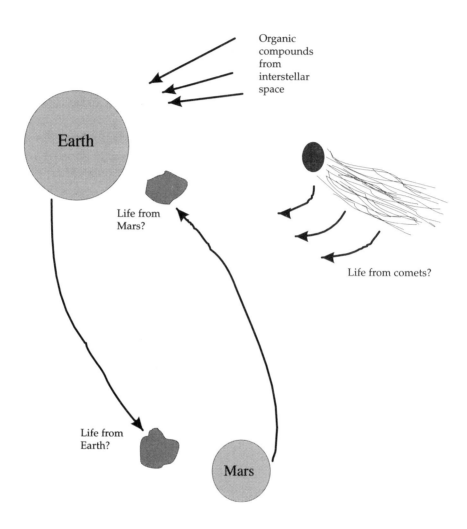

Fig. 14. Some of the basic theories about the origins of life on Earth.

such as the omnipresent PAHs, are known to exist in modern interstellar space. The proposed reactions involving ultraviolet light have also been reproduced in the laboratory, making this part of theory entirely plausible.

Wickramasinghe then proposes that solid particles of dust and ice travelling through space would bump into these free-floating organic molecules, which would collect on their surface. Subsequently the organic coated solid particles would bump into each other and form large clusters of particles within which protective clumps could exist, where unspecified processes would lead to the development of very basic self-reproducing organic walled substances which could be called life. Further aggregations within the early solar system would lead to the formation of planets, where perhaps the primitive life could survive, or comets, where they would continue to survive inside the protective, possibly liquid, centre. As the comet approached the Sun and started to heat up, particles containing these primitive organisms would be left in its trail and would eventually land on other planets, where they would either seed them or, in the case of the viruses, combine with existing organisms to form new varieties of life.

This thesis is a radical one and there are many problems with it, especially regarding the jump from organic molecules collecting in space, for which there is evidence, to the evolution of primitive life, for which there is not. Nonetheless, some of the underlying theories have been taken seriously.

The seeding of planets by passing comets would explain the similarities between the ALH84001 microfossils and those from ancient Earth. Indeed, when the ALH84001 results were announced, Wickramasinghe declared that 'The reason that primitive life on Earth and Mars appear to be similar is that both planets were seeded by similar organisms.' However, there are huge problems with the evolution of life in space, including the extremely low temperatures (approximately $-270°C$), which would slow chemical reactions to almost nothing. There are also massive problems with relating viruses to interstellar bodies such as comets, considering that many viruses have been genetically mapped and related to each other, leaving little room for an exotic origin for any part of them.

Despite the poor reaction to the idea of whole organisms arriving from space, scientific interest does exist in the possibility that large organic molecules could have been delivered into the oceans of newly formed planets.[160] This mechanism could have led to an identical set of organic molecules existing in similar environmental conditions on both Mars and Earth, which could have led to the evolution of identical, bacteria-like organisms on both planets. How likely this is to have occurred is impossible to tell without fossils from the earliest history of both planets.

On Earth different organisms placed in similar environments will often evolve into similar shapes to adapt to those environments. For example, birds, bats, flying fish and insects can all fly using wings, but each is in a separate class of animal and all have developed the ability to fly independently of each other. Perhaps the similar environments on primitive Earth and Mars led to the independent evolution of identical organisms in response to the same environmental needs. Then again, perhaps there is a blueprint which dictates the way organic molecules must combine in order to produce viable organisms. However, we simply do not know, and this first theory of panspermia is very problematic. The second theory of panspermia is probably just as unprovable, but may better explain any similarities between terrestrial and Martian life than cometary seeding.

INTERPLANETARY HITCH-HIKING

The second panspermia theory is based on the premise that life evolved, not in space, but on one planet and then spread through space to other suitable planets. In the case we are looking at here, this would mean that life could have evolved on Mars and then been transported to Earth, or vice versa. Two mechanisms have been proposed which would enable life, or at least microscopic life, to travel successfully between large planets. Both of these are reliant upon meteorite impacts scattering debris from a planet's surface into space. The first proposal is rather reminiscent of the ALH84001 story itself and has been most heavily promoted by H. Melosh.[147]

It is already known that large chunks of rock can be blasted away from a planet's surface by a meteorite impact (see the SNC meteorites in Chapter 11). Melosh, amongst others, proposes that a meteorite impact on a life-inhabited planet could lead to large rocks, containing life, being thrown into space. The cold of space would place any life into suspended animation and in time the chunks of rocks would wander into the path of another planet and survive atmospheric entry to reach its surface. If the planet were suitable (i.e. wet and warm), then the life inside the meteorite would be revived and go forth to seed the planet.

A variation on this theme has been proposed by the Argentinian astronomer Miguel Moreno, who was concerned with the length of time it took large meteors to travel between the Earth and Mars.[147] As the SNC meteorites have shown, it can take anywhere between 600,000 and 16 million years for a large meteorite to travel from Mars to Earth—which is an extremely long time for even the most primitive of organisms to remain alive in suspended animation. Moreno considered these problems and proposed a faster transport mechanism between planets that may give any organisms a better chance of survival.

Moreno's theory is inconveniently called the dust/solar pressure/ keplerian model (DSPK) and again relies on a meteorite impact scattering debris into space. Although this would put larger rock fragments, like those in Melosh's theory, beyond the planet's orbit, it is actually much easier and more common for smaller fragments and particles, less than 1 centimetre in diameter, to be ejected into space. It is just as likely that these smaller particles would contain organisms within them as larger ones. Moreno then used an established theory to explain the rapid transport of these life-packed particles between planets. He proposed that small rock particles could be pushed directly between planets by a mechanism known as solar light pressure. This would make the journey between Earth and Mars, or vice versa, take as little as two months. Even so, could bacteria or similar life survive in deep space for even that long, let alone the thousands of years needed for Melosh's theory?

We have previously discussed the harshness of conditions for life on the Martian surface (Chapter 9); the conditions for life in space are a magnitude worse than that. Even inside the larger fragments, life would have to endure temperatures near to absolute zero and radiation from interstellar space and from their host rock, as well as naked ultraviolet light and zero atmospheric pressure. Under such conditions would any organisms be viable when they reached a suitable host planet?

The answer, according to NASA, is yes—provided the life concerned is shielded from the ultraviolet light. In an experiment, a number of bacterial spores were left in a simulated space environment for periods of up to six years.[91, 92] From this it was found that, if protected from the ultraviolet light, approximately 80 per cent of the spores could be revived after this time. If, however, the ultraviolet light were added, then the survival rate dropped by a factor of four. An experiment performed in space on board NASA's Skylab laboratory produced similar results,[92] again suggesting that bacteria spores could survive for at least several years in space if protected from ultraviolet light. Inside even a small fragment of rock any life would be shielded from the dreaded ultraviolet light and therefore could conceivably survive for a period of time in space. It is, however, doubtful that the bacteria could survive for more than a few decades under such conditions as the desorption of water caused by the vacuum of space gradually leads to irreparable damage of the DNA, killing the cell. Given the shorter length of time involved, Moreno's DSPK theory may therefore be more viable than Melosh's.

Returning to the issue of life on Mars, one or both of these theories could explain a similarity between Martian and terrestrial organisms based on our knowledge of the early solar system. At this time the number of meteorite impacts was many thousands of times greater than it is today, and one only has to look at the surface of the Moon through a pair of binoculars to see the damage it sustained during its early history. An increased number of impacts on a planet's surface would mean an increased amount of debris being blasted into space, and it would only need one viable self-reproducing organism from one planet to seed another successfully. It is easy to imagine an early solar system where literally tonnes of planetary rocks and dust were being batted backwards and forwards between the Earth and Mars like a game of interstellar tennis.

There is still much work to be done on the issue of panspermia, and the only way to prove its existence would be to find life on another planet, or in its fossils, which contained RNA and DNA that could be directly compared to that on Earth. It is not known how or why DNA should have formed to become the means of reproduction of all life on Earth. It is also not known whether the DNA mechanism is restricted to Earth nor whether, through panspermia, we should expect to find it elsewhere in the solar system. If, though, we should find evidence of DNA on Mars, or on another planet, how would we know whether it had evolved separately on that planet or was a relative of DNA on Earth? There are two ways of finding this out.

The first is genetically to map the DNA of the extraterrestrial organism and see whether there are any genes in common with those on Earth organisms. Because all life on Earth has a common ancestor, the genes covering very basic cell functions, such as cell division, can be found in every organism. One would expect some of the same basic genes to occur in any Martian organism that was related to life on Earth.

The second way would be to examine the nature of the genetic molecules themselves. Most biomolecules, including DNA, are capable of being created in two mirror-image forms known as left-handed and right-handed. On Earth DNA is only made of left-handed amino acids (called L-amino acids), and so any extraterrestrial organisms containing L-amino acids are likely to be related to life on Earth whilst those containing right-handed ones (D-amino acids) will not be. Some scientists feel that the question of homochirality (i.e., whether a molecule is left-handed or right-handed) may be our greatest weapon in the quest to identify extraterrestrial life.[14] Indeed, a device called the 'SETH [search for extraterrestrial homochirality] CIGAR ' has been built to help search out and identify any biomolecules in the Martian soil,[134] although there are no plans to use it there yet.

All this theorising would appear to move us back into the realm of Urey's ocean-drenched moon once more (see Chapter 11). However, we ourselves may have already provided a third and much more viable factor to the proposed panspermia theory of interplanetary travel by organisms.

MODERN DAY PANSPERMIA

The following is a theory that has been taken very seriously indeed by NASA, partly because they themselves are the unwitting creators of the transport mechanism by which life from Earth could have reached Mars.

When the Viking landers were in the early stages of design, it was still not fully known what the environmental conditions were like on Mars's surface or, indeed, if there was any indigenous life there. As the landers were to be targeted at areas of the planet that were theoretically the most Earthlike, and therefore the most likely to harbour life, it was decided that a policy must be drawn up to minimise the possibility of life from the Earth travelling with the landers and colonising Mars. As result each part of the Viking landers was rigorously sterilised and decontaminated to remove as many bacteria as possible. Even so, NASA admitted that each lander could still have a microbial population of approximately 110,000 individuals on take-off.[15]

In terrestrial terms this is practically sterile, and NASA themselves estimate that this gives a probability of Mars contamination of 1 in 100,000 from the Viking mission as a whole. The Soviet Union, on the other hand, has never made mention of any decontamination procedures carried out on its Mars landers[99] and may have taken off with microbe levels of many hundreds of thousands per square centimetre. However, the chances of any surviving bacteria being able to revive themselves to reproduce on Mars would still be a remote possibility given the harsh conditions there, especially the ultraviolet light. Nonetheless, we ourselves have enacted the theory of panspermia and, without the aid of either dust or meteorites, have probably exported millions of viable organisms to the planet Mars since the 1970s.

Just in case the panspermia should inadvertently work in reverse (i.e. from Mars to Earth), tens of millions of dollars have been spent building a quarantine facility in preparation for the proposed Mars missions of 2003 and 2005, when it is planned to return Martian rocks to the Earth. A similar procedure was used for the rocks and astronauts returning from the Moon, although the test for micro-organisms in the Moon rocks involved grinding up pieces of rock and feeding them to fish and plants to see if they had an adverse reaction![116] It is hoped that the Martian samples will be treated slightly more imaginatively when they finally arrive on Earth.

PART FOUR

THE MARTIANS ARE COMING!

14. UFOS, MARTIANS AND THE MONUMENTS OF MARS

Aside from the formalised hunt for life on Mars by the scientific community, there is the highly contentious work of astrologers, ufologists and conspiracy theorists, all of whom have reached their own conclusions about what type of life exists or may have existed on Mars. Included in this chapter are the experiences of people who have claimed to have met Martians, either physically or through telepathy, to have been taken to the planet or believe that there is a political conspiracy to cover up known extraterrestrial civilisations on Mars. These experiences and theories, whilst not taken too seriously by astronomers or planetary scientists, do concern the core subject of this book, life on Mars.

EARLY CONTACT WITH MARTIANS

According to Erich von Daniken, our ancient ancestors were meeting, and accepting help from, extraterrestrial beings thousands of years before our modern day speculation about life on other planets. Von Daniken believes traditional cultural stories about gods descending from the sky or arriving in fiery chariots in fact relate to space travellers visiting Earth. Some of these travellers had, according to von Daniken, originated from Mars, making our ancestors the first humans to have a claimed contact with Martians. In *Chariots of the Gods?* he uses ancient Sumerian traditions of approximately 3000BC to suggest that one of its gods, Marduk (= Mars), may have in fact been a visitor from that planet who demonstrated some kind of death ray to the people. Von Daniken writes: 'Time and again Sumerian hymns and prayers mention divine weapons, the form and effect of which must have been completely senseless to the people of those days. A panegyric to Mars says that he made fire rain down and destroyed his enemy with a brilliant lightning flash.'[54] Although von Daniken's books have sold in their millions and received critical acclaim from occultists, the scientific community have not taken them seriously, and it is felt more probable that the Sumerians were describing a volcanic eruption or something similar rather than a Martian death ray.

It is necessary to jump from 3000BC to the seventeenth century AD before the next contact with Martians can be found. This case is the first of a number of examples where the Martians supposedly contact individuals using telepathy. Emmanual Swedenborg was a Swedish polymath who was born in 1688. He was an outstanding scholar for his time and wrote books on a whole variety of scientific subjects, including astronomy, mathematics and engineering. He designed buildings, wrote music and could speak several languages, and he was a popular and famous member of the Swedish aristocracy. There was, however, an event that was to steer him away from scientific study and into the murky world of the occult and religious philosophy. In 1744, after a particularly hearty meal, Swedenborg experienced a highly lucid vision in which he was taken to see heaven and hell and even met Christ. From this moment on he claimed to have been blessed with extrasensory perception and is credited with a number of remarkable predictions and similar psychic experiences. He also claimed to be living half in and half out of the world of the dead and to be able to travel spiritually wherever he wished. Included in these journeys were visits to other planets within the solar system.

Whilst on the planets Swedenborg met many of their inhabitants, and gave detailed descriptions of their appearance and behaviour. He described people on the moon as being '. . . small, like children of 6 or 7 years old, but [having] the strength of men like ourselves. Their voice is like thunder and comes from the belly for the Moon is in quite a different atmosphere from the other planets.'[42] The Martians, however, were highly revered by Swedenborg, who describes them as 'the best of all spirits in the planetary system. Their gentle, tender, zephyr-like language, is more perfect, purer and richer in thought, and nearer to the language of the angels, than others. These people associate together, and judge each other by the physiognomy, which amongst them is always the expression of their thoughts. They honour the Lord as sole God, who appears sometimes on their earth.'

Shortly after Swedenborg's time lived the British author Jonathan Swift, whose most famous work was the adventure story *Gulliver's Travels*. In recent times it has been claimed that Swift might have had contact with extraterrestrial intelligence, probably Martian in origin, based on the following extract from his famous work:

'The Laputan astronomers . . . have discovered two lesser stars, or satellites, which revolve about Mars, whereof the innermost is distant from the centre of the primary planet exactly three of the diameters, and the outermost five; the former revolves in the space of ten hours, and the later in twenty-one and [a] half; so that the squares of their periodic times are very near in the same proportion with the cubes of their distance from the centre of Mars, which evidently shows them to be governed by the same law of gravitation that influences other heavenly bodies.'[224]

This is an uncannily accurate description of Phobos and Deimos, Mars's two moons, which were not to be discovered until 1877, 151 years after the publication of *Gulliver's Travels*. This has been used by modern UFO researchers to suggest that Swift had been contacted from space. The truth is, however, probably more mundane.

Swift was a keen astronomer and would have been versed in two important theories at the time regarding Mars's moons. The first of these, by Tycho Brahe, stated that Mars must have two moons because Venus had none, Earth one and Jupiter four. In other words, the number of moons doubles with increasing distance from the sun. A second, and more scientific theory, by Johannes Kepler used orbital variations to suggest that two moons were present. It is reasonably certain that it was upon Kepler's work that Swift based his detailed description of the Martian moons and not on personal description from aliens.

We have to move forward considerably in time before the next contact occurred with the inhabitants with the Red Planet.

PSYCHIC CONTACT IN THE 1890S

The 1890s seems to have been something of a landmark in terms of psychic contact with Martian intelligence for, in that decade, no fewer than five spiritualist mediums claimed contact with the planet's inhabitants. The first, and best known, of these contacts occurred in 1894 to a Swiss medium named Helene Smith (her real name was Catherine Elise Müller), who lived in Geneva. A person present at a seance given by her was so impressed by what he saw that he asked the eminent psychologist Theodore Flournoy to examine Helene. Flournoy was extremely sceptical about the whole spiritualist movement and was vocal about it to Helene. Nonetheless, the two of them seemed to hit it off together and he agreed to study her from a purely psychological perspective.

Flournoy used hypnosis to place Helene into deep trances, during which she claimed to contact dead people such as Victor Hugo and to relive a number of past lives, including one where she claimed to have been a Hindu princess and Marie Antoinette. Flournoy soon became fascinated with one of her more unusual claims—that she was in regular contact with Martian beings. Helene would have visions where she would see herself standing on the surface of Mars interacting with the beings that she met there. For example, in 1896 she described sitting on a Martian bench next to a pink lake over which ran a bridge constructed from yellow tubes whilst all around were Martians, some of whom were capable of flying. Her other descriptions of Mars were of a planet with a yellow sky and flowing water, with beings that were identical to humans except that both sexes wore an identical uniform. The beings travelled using horseless carriages and lived in strange box-like bungalows with fountains on their roofs. Most interestingly to Flournoy, Helene claimed to be able to speak and write the Martian language.

The language was a strange one that sounds remarkably like French in its structure and vocabulary. The Martian script itself looks more like a cross between the written forms of Arabic and Russian. Helene was consistent in her use of this language, and an example of it was provided by her to accompany a drawing of a Martian house she had just made. The caption to the picture was 'Dode ne ci haudan te meche metiche astane ke de me veche', which meant that the house she had just drawn belonged to Astane.

Flournoy's analysis of Helene, published in a book entitled *From India to the Planet Mars*, was a sceptical and critical one.[70] He deduced that Helene believed she was born for better things and that her fantasies of past lives and Martian civilisations were a mental attempt at trying to escape from her humdrum background. Her past-life knowledge was put down to cryptomnesia (suppressed memories) and the Martian episodes to flights of fancy. More recently, some authors have expressed a belief in Helene's power[104, 243, 244] as a medium but not in her contact with Martians.

In 1895, the year after Helene Smith's first contact, another medium, called simply Mrs Smead, the wife of a clergyman, was also claiming to talk to Martians. It is unlikely that this case was linked to that of Helene Smith as Mrs Smead was an American and as Flournoy was not to publish his results until four years later. In the same manner as Flournoy investigated Helene, so another psychologist named Professor Hyslop undertook to examine Mrs Smead.

Mrs Smead's experiences began when she claimed to be able to contact the departed spirits of her three children and brother-in-law. These spirits claimed that the souls of the dead went to other planets in the solar system and that the brother-in-law and her daughter were on Mars whilst another daughter was on Jupiter, which was referred to as 'the babies' heaven.'[102]

Her daughter, named Maude, gave details of Mars which are essentially Lowellian in nature, with descriptions of canals, dying civilisations and vegetation. The Martians were described as being very similar to humans, and their houses, clothes and transport were variations upon our own. Interestingly, all Martian men were forced to work in the fields until they married. Hyslop used another popular esoteric phenomenon, multiple personality disorder, to explain Mrs Smead's contact with her dead extraterrestrial relatives. Her believed that she had a hidden secondary personality that was feeding her first personality with the visions of Mars.

It is interesting to note that, prior to Flournoy's and Hyslop's experiences, Camille Flammarion, the noted astronomer, had seriously proposed trying to communicate telepathically with the Martians. A tongue-in-cheek editorial in the *Independent* periodical of 1909 discussed the above two cases, concluding that 'The two descriptions do not agree at all in architecture, costume and language, but there is nothing in that to discourage the psychic reader. The two mediums may have been seeing different parts of the planet.' [103]

The planet was again contacted in 1895 by a middle-aged French woman known by the pseudonym 'Mireille'. Her ability to converse with aliens was discovered when she was hypnotised by a friend of hers, Colonel de Rochas, in order to relieve her of a painful ailment. De Rochas was interested in researching the phenomenon and continued to hypnotise her for some time after this. As she was placed into deeper and deeper trances, her descriptions and conversations become more fantastic.

On one occasion she reported meeting a friend of hers called 'Victor', who had been dead for a number of years. Victor was apparently happily living on Mars and was eventually to 'penetrate the electrical field' surrounding Mireille so that he could communicate directly through her with de Rochas. We can learn little of Mars from Victor's rather vague descriptions, except that there were canals on the surface and that the Martians were similar to humans except for massively long arms which were used as 'organs of affection'. Unfortunately, de Rochas seems to have been more interested in debating the philosophy of personality survival with Victor than in Mars and spent a number of weeks trying to convince Victor that he was a product of Mireille's imagination! This was also to be his final conclusion on the case.[189]

The next case was to fall under the gaze of the famed Swiss psychologist Carl Jung who, in 1899, came across a 15-year-old girl known to us only as 'S.W.' Jung describes her as being normal apart from being prone to occasional trance-like states in which she would see the spirits of the dead (something that I would describe as being slightly abnormal). During these episodes S.W. would leave the Earth and travel between the stars, which she claimed were filled with a host of different spirit worlds. Mars was one of these inhabited spirit worlds, but she mentions it only briefly as being 'far more advanced than we are. Thus, flying machines have long been in existence on Mars; the whole of Mars is covered with canals; the canals are artificial lakes and are used for irrigation.'[109] In common with the other psychologists, Jung became convinced that S.W. was suffering from a dissociated personality, something with which she strongly disagreed. After her brief flirtations with spiritualism, S.W. abandoned it in favour of more conventional work in an office.

The fifth and final medium contact with Mars has few details attached to it. Some time around the turn of the century the occultist Lewis Spence placed an unnamed medium into a deep trance using magnets and was surprised to find that she was describing scenes from Mars. The description of the medium's vision is absolutely minimal, Spence saying that she 'found only images of fright and horror.'[220] This generally contrasts with the descriptions of the other mediums contacting Mars from this time.

We must jump forward a few years to find the human race's next contact with the Martians. This time it was to be a face-to-face meeting as opposed to the previous telepathic encounters.

UFOS AND THEIR OCCUPANTS

From the moment Kenneth Arnold reported seeing his 'flying saucers' in 1947, the notion of extraterrestials changed from their being the passive inhabitants of interstellar worlds and instead were now marauding armies of violent creatures capable of building their own spaceships and invading Earth. When the UFO craze took off in the 1950s Lowellian Mars was in a steep decline. Ufologists and science fiction writers (see Chapter 15) had breathed new life into the possibility of intelligent civilisations on other planets, especially Mars, and the phrases 'Martian', 'Men from Mars' and 'Little Green Men' became household terms that are still in use today. In this pre-space age environment, reported meetings with Martians were a reasonably common experience, and some of these are discussed below.

The initiator of contacts with alien extraterrestrials was a man called George Adamski, a Californian travelling wine salesman of Greek origin. In 1953 he wrote a best-selling book, entitled *Flying Saucers Have Landed!*,[3] in which he details his contacts with beings from other planets in the solar system. These beings chiefly came from Venus, but he also had encounters with those from Mars and Saturn. He claimed that in 1946 he had encountered a flying saucer in the Californian desert and after years of trying to contact the aliens had been invited aboard another craft that he had come across in 1952. On this and other occasions Adamski was taken across the solar system by his Venusian friends and shown that all the major planets, including the Moon and Mars, had atmospheres, equable climates and life. His descriptions of life on Mars are vague, although the aliens themselves are described as being almost identical to humans. Adamski's books were international best-sellers, and although Venus was Adamski's main concern, it was he who started the whole craze for extraterrestrial encounters of the third kind.

At around the same time as Adamski was gaining publicity, a photograph was released purporting to show a Martian captured by the military outside Mexico City in 1950. The photograph itself is now a UFO classic, depicting two sinister men in military-style raincoats and wide-rimmed hats holding between them a naked and skinny thigh-high figure. Strangely, two old peasant women can be seen in the background, apparently staring at the Martian. This photograph has been analysed many times over the years and no firm conclusions have been reached. There were suggestions, based on an earlier hoax, that the figure was a shaven monkey, but this is generally thought not to be the case. Instead, it is likely that the photograph has been doctored in some way, with the Martian painted in over a perfectly normal object. This may have been a pram, a bicycle or a similar handlebarred object as the angle of the men's grips suggests that they are both holding on to a horizontal tube.

In 1954, the year after Adamski's first book was published, a gentle Englishman named George King was enjoying a quiet Saturday afternoon when a voice in his head said to him, 'You are to become the voice of the interplanetary parliament.' The encounter caused him to form the Aetherius Society and to claim to be the Earth's mouthpiece for the parliament. King headed the society and held regular meetings, during which he would receive instructions from the Master Aetherius, who, although a Venusian, was broadcasting from 'Mars, sector six'. The society was surprisingly popular, and before long King was claimed to have met with Jesus Christ, Lord Buddha and St Peter, all of whom were avowed Aetherians. Mars, apart from being the Master Aetherius's base, features little in the society's workings, which are more concerned with the politics of the interplanetary council. The council's transactions are unusual.

For example, in the 1960s there were continual warnings that the Saturn council was not prepared to let man land on the Moon and that any attempt to do so would meet with disaster. Best of all is a recent warning by the parliament that an underwater race of fish on the planet Garouche are trying to kill all land-based life on Earth by sucking away the air in the atmosphere. Apparently the marine life here will be spared because the oxygen in water does not come from the atmosphere, but from 'an unknown source'. The society is still around today and I would recommend attending some of its talks—though probably not for their factual content.

Another early, and unusual, physical encounter with Martians comes from Italy in 1954. On 1 November that year a 40-year-old peasant woman named Rosa Lotti awoke early to go to church. In order to get there she went through a wooded area, through which she carried her shoes, so as not to soil them, and a bunch of flowers. As she passed through the trees she spotted a strange object, to which she paid little attention. Suddenly she found her path blocked by two strange men who were human-like in every detail apart from being barely one metre tall. The men snatched her shoes and flowers away and placed them inside the strange object. At this point Rosa made a hasty exit and went immediately to church, where she told the story to her priest. It was he who suggested that she had met with 'men from Mars' and that the strange object must have been a flying saucer. The Martians were never seen again and, although the woman stuck to her story, the family received much ridicule after the publication of the story in the national press.[64]

Earlier in that same year another encounter with a Martian took place near Lossiemouth, Scotland. On 18 February one Cedric Allingham was out bird-watching when he saw and photographed a UFO in the cloudy sky. He must have been spotted by the aliens for five minutes later the craft, described as being 18 metres in diameter with a domed top and flat bottom, landed not 50 metres from where he stood. Allingham approached the craft and as he did so a door opened and a smartly dressed, two-metre-high man stepped out. The UFO's pilot then proceeded to have a discussion with Allingham, during which time he claimed to have come from Mars and was wistful about the lack of water there and the poor state of their canals. After a short time the pilot returned to his craft and sped off into the sky once more. Allingham wrote up his experience into a book which was unimaginatively entitled *Flying Saucers from Mars.* Although it was later claimed that he disappeared mysteriously during a lecture tour of Sweden, it should be noted that Cedric Allingham was actually the pseudonym of a highly successful novelist.[5]

An almost identical experience occurred to the American farmer Gary Wilcox, who came across a silver UFO in one of his fields in 1964.[231] Beings from the craft were busy collecting soil samples when he approached them, but they seemed willing enough to talk to him. They told him that they were stopping off at Earth after a soil-collecting mission to Mars and that, in their opinion, Mars would one day replace Earth in terms of importance in the solar system. After requesting a sample of his nitrate fertiliser, the beings boarded their craft and left.

This is one of the last reported physical contacts with Martians for, as unmanned spacecraft sent back definite information about Mars's hostile environment, so the probability of civilisations living there disappeared from science fiction films and books and therefore also from the public's eye. It seems that to have an extraterrestrial visitor come from Mars, or indeed any of our nearby planets, was to risk an unnecessary confrontation with planetary scientists, so from the mid-1960s onwards aliens came from other stars or just refused to disclose their origins. In fact, it is very difficult to find any reference to self-confessed Martians visiting people from this time. The term 'Martian' was, and still is, attached to some cases but only as a descriptive word for extraterrestrials rather than because they came from Mars. This does not mean that the idea of Martian civilisations were not being reported or discussed: just because they were not visiting us, it did not mean that they were not there on the planet hiding from us or waiting for us to contact them.

In the 1970s there was a radical change in the style of reported encounters with UFOs. Instead of mysterious or friendly encounters with silver, cigar-shaped objects, aliens suddenly became sinister entities who travelled in dark triangular craft abducting people to carry out personal medical experiments on them. Another culture grew around this new style of alien—that of the conspiracy theory.

CONSPIRACY THEORIES

Over the centuries many organisations, religions and individuals have been accused of conspiring against society. For example, the Germans blamed the Jewish community for many of their country's pre-war problems and, similarly, the Freemasons, Cathars, Catholics and Protestants have all been persecuted for apparently organising anti-society activities. In the past it has normally been governments that have accused sections of society or organisations of conspiring against the state, but in more modern times this trend has reversed, so that it is now the people accusing governments of conspiring to cover up vital information regarding the political future of the Earth.

Conspiracy theories in the form discussed here are a recent, and mainly American, phenomenon. After the fanatical McCarthy communist hunts of the 1950s, the more liberal 1960s and the Vietnam War allowed society to question the motives of its leaders. The first major government conspiracy theory was the supposed CIA involvement in the assassination of President John F. Kennedy, and after the Watergate affair people suddenly realised that their governments did not always work for their best interests. After this fringe groups with their own anti-government conspiracy theories were suddenly being taken seriously.

A great number of modern theories are centred around accusations that governments have been concealing contacts that they have had with UFOs. These are numerous and range from the American government's recovery of a UFO at Roswell in 1947 to accusations that governments help aliens kidnap people for use in extraterrestrial medical experiments.

The chief conspiracy theory associated with Mars has caused the biggest wave of popular interest in the planet since Lowell's canals. Indeed, there are many parallels between it and the whole canal issue, with a factual outcome that will probably be similar to that of Lowell's vision of Mars. The case in question has become known as 'The Face on Mars' and is a fascinating one.

THE FACE ON MARS

In vogue with other modern conspiracy theories, the Face on Mars started as a very small snowball that has avalanched into an all-encompassing vision of an extraterrestrial civilisation on Mars.

The incident originated in the summer of 1976 when the Viking 1 orbiter craft took close-up photographs of the Cydonia region in the northern subpolar hemisphere of Mars. When the photographs had been beamed backed to Earth and cleaned up they were examined closely for any interesting geological or geographical features. During this examination a NASA scientist noticed a formation on the ground that resembled a human face and drew it to the attention of colleagues. At the time NASA was keen to keep its Viking publicity machine going and thought that the photograph would make an interesting feature for newspapers and magazine to print. The media duly responded and the article was published in *Science News* on 7 August 1976.[211]

The article was accompanied by a rather grainy and contrasty photograph of a 21 by 13 kilometre section of the Cydonia region taken from a height of 2,090 kilometres. In the upper right of this photograph there is a raised feature, half in shadow, that strongly resembles a face complete with a forehead, eye, nostril and mouth. It also appears to have some form of mantle around it that looks like an Egyptian headdress.

Almost as soon as the photograph had been published, UFO researchers started to comment on the possibility that the feature could have been artificially built by extraterrestrials. This was reinforced by the discovery of a pyramid-shaped form to the left of the face which, with its Egyptian features, was soon linked into the story. NASA acted swiftly and, perhaps to stifle the growing adverse publicity, announced that a second image taken 'several hours later' under different light conditions revealed that the features were natural formations and were an 'oddity of light and shading'. This seemed to satisfy most people and the issue died down.

In the early 1980s two engineers, Vincent DiPietro and Gregory Molenaar, were working at NASA's Goddard Space Flight Center when they chanced upon the original photographic frame containing the Face and were extremely impressed by its likeness. They began conducting their own research into the case and before too long unearthed a second Viking photograph of the face which had apparently been misfiled by NASA. This second photograph had not been taken several hours later, as NASA had claimed, but in fact several days and 35 orbits later. The sun angle differed slightly from the original photograph but the face was still clearly visible on the Cydonia plain. After computer enhancement the face stood out even more and prompted the two engineers to write a book about the affair in 1982. The book, entitled *Unusual Mars Surface Features*, suggested that the face was of artificial origin and noted possible links to Ancient Egypt.[59] It did not, however, receive widespread publicity and reviews were generally negative. Nonetheless the pair continued their research and even promoted their ideas in lecture tours across America.

A second book by R. Pozos, entitled *The Face on Mars: Evidence for a Lost Civilisation?*, was published in 1986. It made similar suggestions to those by DiPietro and Molenaar but again received limited coverage.[170] The big step forward for the whole issue came in 1987 when Richard Hoagland published his book on the affair, romantically entitled *The Monuments of Mars: A City on the Edge of Forever*.[87]

Hoagland was a former science writer and NASA consultant who had founded a research organisation called the Mars Mission (TMM) whose purpose was to study objectively the Face on Mars and its associated monuments to see if there was any possibility of an artificial intelligence having built them. His 1987 book represented the fruits of this research, which, judging by some of its conclusions, may not have been as entirely objective as he may have claimed.

Hoagland's quoted reaction on first seeing the enhanced face on Mars photographs was, 'Either this was the most incredible thing I'd ever seen and had to be pursued, or it was a complete waste of time. If it was real then the benefits to mankind would be off scale; if a waste of time, then it wouldn't take long to figure out.' Despite this apparently open-minded attitude, he seems to have started his research on the assumption that the face is of artificial construction, although there are other possible reasons for its shape (see below).

His first step was to search around the face to try to find other, similar artificial landforms. The face itself lies on a relatively flat, featureless plain, but there is a group of mountains and hills about 18 kilometres to the southeast of it. Examining this area in detail, Hoagland found another series of artificial landforms which could be linked to the face. This includes a group of polyhedral features which were collectively called 'The City', a large polyhedral pyramid named the 'DiPietro and Molenaar (D & M) Pyramid', a circular mound with a spiral marking running round it and a ditch around the base named 'The Tholus' and a narrow, flat-topped ridge known as 'The Cliff'. To Hoagland all these features were part of a complex city network built by an extraterrestrial civilisation, and to prove it he turned to the field of mathematics.

The impetus behind this was to try and find alignments between key points in the city complex and astronomical features, such as are found at Stonehenge and other ancient sites. Such an alignment could only be found by calculating the position of the Sun and stars back in time until, 500,000 years ago, the sunrise on the summer solstice aligned the eyes of the face with the central square of the City. This, said Hoagland, meant that the whole complex was built half a million years ago.

In conducting this exercise he noted that if two separate lines were drawn from the City so that they passed through the eyes and chin of the Face, then they would converge at the cliff. Inspired by this, he drew connecting lines joining every object to every part of every other object and measured the angles made between them. He discovered that the pattern produced was a complex geometric one that would be hard to obtain by chance and that many of the angles measured contained mathematical constants such as pi and the base of natural logarithms. On top of the mathematical constants, the angle 19.5° seemed to be significant within the landforms themselves, recurring again and again. Through tetrahedral geometry, Hoagland thought that the 19.5° referred to the latitude of Mars and duly found a string of volcanoes there (only Olympus Mons is actually on this latitude). Looking to the other planets, he found that on Earth the Hawaiian volcanoes were on this latitude and that Jupiter's Red Spot was as well.

The location of so many geographical features in the same position on different planets was significant to Hoagland, who felt that the aliens were trying to tell us something. Although it is not clear how, Hoagland felt that the complex mathematical code of the Cydonia cities was indicating that a new form of energy could be gained from 'the way spinning bodies interact with and control gravity by the creation of a vortex that "sucks" in "free energy" from a hyperdimensional level'. This would solve all the Earth's energy problems in one go. At the same time Hoagland also became convinced that the builders of Cydonia were not bone fide Martians, as Mars is too inhospitable for life to have evolved to any level, but probably visitors from another solar system who were just passing through. If they stopped at Mars half a million years ago then they must have also stopped at Earth—but is there any evidence of their having been here? The Mars Mission decided to investigate.

Eventually, a scenario similar to that of Arthur C. Clarke's *2001: A Space Odyssey* was envisaged wherein the aliens visited our ancestors, helped them to evolve into us and then left an encoded message on the Martian landscape so that we would only find it when we were advanced enough to understand its meaning.

Convinced that some remains of the extraterrestrials' influence on our ancient ancestors would be preserved in their artefacts, Hoagland and his followers set about comparing the Cydonia monuments to prehistoric relics on Earth. They were spurred on by the work of Russian geologist Vladimir Avinsky, who thought that the Face was 'the Martian Sphinx', and Ancient Egypt was examined first. Avinsky believed that the Face itself could be a Martian version of the Sphinx and that the tetrahedral landforms, such as the City, could be the equivalent of the Pyramids. Hoagland measured the geodetic latitude (one where the curvature of the Earth is taken into account) of the Egyptian Sphinx and found that its cosine equalled that of the D & M Pyramid on Mars. With the link made between Egypt and Cydonia, the Mars Mission then used unproven work by the geologist Robert Shoch to suggest that the Sphinx may have been built before the last Ice Age and, as no civilisation on Earth was advanced enough to construct it then, implied that it may have been of extraterrestrial origin. A British member of the Mars Mission then drew parallels between the ancient stone circle at Avebury in southern England and the features on Cydonia. By placing Silbury Hill (a man-made mound) on top of the Tholus, a crater to the left of the face coincides with the village of Avebury itself.[153]

The Mars Mission is still very active and has begun lobbying the United Nations and other organisations to examine its data. Interest has been thin so far. Other people have looked at their data and come to their own conclusions.

One of the biggest pieces of research done on the Cydonia monuments was that by Mark Carlotto, who used computer-imaging techniques to clean up and analyse the Viking photographs. In 1991 he produced a book on the subject (*The Martian Enigmas—A Closer Look*), which was to take Hoagland's Mars Mission work a stage further.[37]

Firstly, Carlotto greatly enhanced the clarity, focus and sharpness of the rather fuzzy and blurred Viking pictures, revealing a large number of previously unseen features on the plain including, according to Carlotto, the presence of an eyeball and teeth in the Face. Carlotto also used three-dimensional imaging to produce oblique reconstructions of the landforms so that they could be viewed from different angles. Secondly, he used fractal analysis on the Viking images. This is a statistical technique, used extensively in the Gulf War, that is designed to identify symmetrical objects hidden in the landscape, for example tanks or aircraft camouflaged in a desert, on the theory that they may be man-made. Using this analysis, he identified the Face as being the most symmetrical object, and therefore possibly artificial, in a 4,000 square mile region!

All the interest that was being shown in the Cydonia region would, it was hoped, be proved one way or the other as the Mars Observer probe (see Chapter 8) neared the planet in the summer of 1993. All those involved in the Cydonia analysis had been lobbying NASA for years about giving the region priority when the probe went into Martian orbit. NASA consistently refused to do this, instead saying that they would get around to photographing the region in good time. The pro-Cydonians then requested that they should have instant access to any photographs produced by the Observer. Again NASA refused, saying that it could be up to six months before the photographs were available to the general public. It did, however, say that anybody could come along to Pasadena to watch the photographs come in live on screen, but, for scientific and copyright reasons, no actual hard copies would be available immediately.

Despite this offer, rumours of a cover-up began to circulate through the UFO community and suddenly NASA found itself at the centre of a growing controversy surrounding the Cydonia Face. It did not make things better for itself by releasing statements that still insisted that the features were natural landforms viewed under strange lighting conditions. Ufologists began to suspect that NASA had something to hide in respect to the Cydonia region: a conspiracy theory had started to grow and was about to explode in NASA's face.

The event that was to give the theory substance occurred on 21 August 1993 when, shortly before it entered Mars's orbit, all contact was lost with the Mars Observer probe. The conspiracists became convinced that the probe had either been 'turned off' or was still secretly working so that NASA did not have to reveal the truth about the past existence of extraterrestrials on Mars. Hoagland's group picketed NASA, accusing the organisation of a 'billion dollar cover-up'. He became convinced that the probe was still transmitting and lobbied the Jodrell Bank radio telescope to search for its transmissions. Jodrell Bank was looking for the Observer anyway but reported that no trace of it could be found. At the same time the National New Age, Alien Agenda and Cosmic Conspiracies Conference in Phoenix collectively directed 'love energy' at the probe to make it well again. Finally, a note, made of cut out newspaper letters, was delivered to the Jet Propulsion Laboratory which read 'WE HAVE YOUR SATELLITE IF YOU WANT IT BACK SEND 20 BILLION IN MARTIAN MONEY. NO FUNNY BUSINESS OR YOU WILL NEVER

SEE IT AGAIN.' Another promoter of the NASA conspiracy theory is an academic philosopher named Stanley McDaniel, whose book on the subject, *The McDaniel Report*, makes some recommendations to NASA concerning future Mars missions (none of which, as far as I know, has been adopted).[139]

The arrival of the Mars Global Surveyor in 1997 has brought these questions back into focus, and again NASA is at the centre of the Cydonia conspiracy theory. Fortunately, influential as the Mars Mission and its followers think they are, NASA has learned to ignore their lobbying. On 5 April 1998 the Global Surveyor probe took an enhanced photograph of the Cydonia region. Although the probe was over a year from its closest orbit, the photograph revealed the Face to be an eroded hill. Face fans, however, still maintain that the feature is artificial and have published their own processed version of the NASA images.

CONCLUSIONS ABOUT THE FACE

Few people amongst the scientific community take Hoagland *et al* very seriously, instead believing that the Face and its associated landforms are nothing but Martian examples of simulacra (natural objects which resemble biological or man-made ones). Ironically, Carlotto's enhanced photographs go a long way towards proving what exactly the Face and Pyramids actually are.

Beause of the low angle of the sun at the time of the photography, the enigmatic Face is half in the shade, obscuring its right-hand side. Carlotto's enhancements partially remove the shadow, revealing not a symmetrical feature, but that the right-hand side of the Face is missing. Without the shadow the face more resembles an eroded hill. The reconstructed oblique views that Carlotto provides show the Face dissolving into a series of low, gentle hills, with the 'mouth' region becoming a narrow dry valley on top of a central ridge. A similar effect can be seen on the Pyramids, which, when viewed closely and from different angles, reveal themselves to have crooked margins and bifurcating ridges and may not result in perfect tetrahedral landforms, as depicted by Hoagland's mathematics, but instead ragged features that closely resemble the wind-faceted hills seen in many of Earth's driest deserts. As for Hoagland's geometric measurements, anybody can find significance by drawing lines between landforms to produce a series of triangles: the latter are geometric objects that are bound, by their trigonomic nature, to have mathematical significance within them.

The Cydonia conspiracy theory has come to fruition with the arrival on the scene of remote viewers. The latter are psychics who claim to be able to travel to distant locations using their minds. The remote viewer Joe McMoneagle visited the Cydonia region and described finding 'a very large pyramid with interconnecting corridors and very large rooms. There were at least four or five [pyramids] laid out in a specific geometric pattern . . . towards the end of the session I became aware . . . of who had constructed the pyramids. I had the perception that there were a race of beings or humanoids that were sort of passing through at the time. My sense was that they were moving through the solar system and had to move on to a different location. Its possible they moved here!'[162] In the same way that the mediums of the 1890s were contacting beings who were living on a Lowellian Mars with canals and vegetation, so the modern remote viewers were seeing Hoagland's vision of Mars with its transient aliens creating meaningful geometric patterns out of their pyramids.

It is worth mentioning some of the other analogies to the Face, which include comparisons to Elvis Presley, the Cybermen from *Dr Who* and an ape. One British astronomer, Eric Crew, has done work similar to Hoagland's and believes, by connecting lines from the Face to nearby landforms,[51] he has found a map of our solar system on the Cydonia plain. We have not heard the last from the Cydonians.

THE PATHFINDER MISSION

Unsurprisingly, within a week of the Pathfinder probe landing on Mars there were a whole new set of conspiracy theories put forward.[187] Initial reports said that the Pathfinder pictures of the Martian landscape were in fact of a movie set in Arizona and another that certain photographs were being withheld by NASA because they showed a mineral water bottle in them! In addition, there were claims that once the mission fulfilled its real, but hidden, objective to find life, the cameras would go mysteriously dead and the information be kept secret. Indeed, the chief camera operator of the mission, Peter Smith, is quoted as remarking about the range of objects supposedly spotted in the landscape photographs, 'You name it, they're seeing it!'. He also added a comment about the outline of a sun god having been reported in a Martian rock: 'They outlined its pattern in a rock and, by God, once they did that you couldn't help but see it!'[187] Most bizarre of all were three Yemeni men who have tried to sue NASA for trespassing on Mars 'without informing us or seeking our approval'. The men have claimed that Mars was given to their ancestors over 3,000 years ago. The lawsuit was withdrawn after the three were arrested for time-wasting.

OTHER SIGNS OF LIFE

Although the Cydonia Face is the subject of the most popular alien theory about Mars since Lowell's canals, there have been more minor signs of life seen over the years. Here are some of them.

The second mostly widely held theory about Mars is that one of its moons, Phobos, is an artificial satellite. This was initially proposed in 1882 by Mahatma Kuthuami Singh, who commented that 'The inner satellite, Phobos, is no satellite at all. It keeps too short a periodic time.' Since then this theory, based on the small size of the moon and its irregular orbit, which suggests that it is lighter than it should be, has been proposed by a large number of people, including George Adamski and Carl Sagan. The Russian Phobos missions of 1988 (see Chapter 8) showed photographs of the moon, but there was no evidence of artificial life on it. However, ufologists have blamed the failure of the Phobos missions on extraterrestrial interference and claim that the final picture taken by Phobos II shows an artificial object hurtling towards it on a collision course.

Similarly, the *Los Angeles Times* of 28 January 1950 reported that the Japanese astronomers Sadao Saeki had seen a huge explosion on Mars on 16 January which produced a mushroom cloud 1,450 kilometres in diameter. Saeki said that the cloud 'was like the terrific explosion of a volcano' and that it would have resulted in a catastrophe for 'any form of Martian life'.[194] No other people observed it, but it has since been speculated that the explosion may have been Martians testing their own atom bomb![19]

Other signs of life on the Red Planet include theories that people from Earth are being kidnapped as Martian slaves[72] and that the Martians have built a huge dome around the outside of the planet to hide their canals and cities from our visiting spacecraft.[231] Other people have reported seeing lights on Mars, having obtained silk from Mars and that the manna of the Israelites was of Martian origin. It was also suggested that the mysterious green children of Woolpit, a nineteenth century mystery involving the spontaneous appearance of two children in a village, were Martians whose skin was adapted to the lesser strength of the Sun there.[19]

The willingness of people to believe in the possibility of Martians was revealed in 1978 when an elaborate April Fool's joke, produced by Anglia Television in Britain, was believed by thousands. The programme, called *Alternative 3*, painted a doomsday scenario for Earth and claimed, amongst other things, that manned bases were in operation on both the Moon and Mars and that our most important people were emigrating there before the Earth's imminent destruction. The documentary style production had people on both sides of the Atlantic fooled, many of who still believe in the underlying truth of *Alternative 3*.

In this chapter we have seen how Mars is still a fertile ground for ufologists and conspiracists. In the next chapter we will see how big government has used the Martians as a means of portraying their own political paranoia to the public at large.

15. THE WAR OF THE WORLDS

One of the most famous incidents connected with the 'life on Mars' issue occurred on 30 October 1938 when thousands of Americans heard an early evening radio weather forecast apparently interrupted by a newsflash declaring that 'a huge flaming object' had crashed to Earth and that alien machines had emerged firing deadly death rays at anything that moved. Absolute panic followed this announcement, with people running from their homes carrying guns, axes, knives or any other weapon with which they could defend themselves against the incoming Martian invasion. The churches were filled with people praying for divine help against the beings from another planet, whilst others barricaded themselves inside their homes. As far as many people were concerned, the end of the world had come.

Fortunately for these hysterical masses, the radio broadcast turned out to be the first part of Orson Welles's adaptation of H. G. Wells's classic science fiction book *The War of the Worlds*.[241] Although he was heavily criticised for the panic he created, Welles proved at first hand that belief in the possibility of a Martian invasion was very strong in most people's minds and that this fear could be tapped into with great ease. The style and format of that 1938 radio play was to be repeated again and again over the coming decades, the Martians themselves becoming a subliminal expression of our own, and our politicians', greatest fears. Before we delve into the world of the science fiction portrayal of Martians, it is worth mentioning that it is still possible for dramas to cause panics along the same lines as Welles's.

In 1978 the *Alternative 3* drama (see Chapter 14) caused a wave of paranoia that can still be found amongst conspiracy theorists and ufologists. On another level, the BBC broadcast a play called *Ghostwatch* on Hallowe'en in 1993 but, in the same vein as Welles, did not announce it as a drama beforehand. The programme involved apparently live reports from a poltergeist-afflicted council house in England. As the night drew on the public were first convinced that the case was a fraud, only to be suddenly confronted by an apparently genuine burst of violent poltergeist activity that resulted in the abrupt termination of the broadcast. People jammed the BBC's switchboard in blind panic at what they had witnessed, only to be told that it was just fiction. Outrage followed over the ensuing days as people accused the organisation of deliberately setting out to cause panic, and the suicide of one person was blamed on the programme. It is our willingness to believe in the possibility of poltergeists and Martians, as well as our desire to be scared by them, that has led to the plethora of science fiction books and films dealing with such topics, especially alien life forms—including the Martians.

The book which Welles's production adapted, *The War of the Worlds*, represented the first mass popularisation of Martian life, and its author, H. G. Wells, was clearly heavily influenced by Percival Lowell's vision of a desiccated Mars inhabited by a dying civilisation. *The War of the Worlds* was published four years after Lowell's first writings on the subject of the canals of Mars, and its opening lines could have been written by Lowell himself:

> The secular cooling that must some day overtake our planet has already gone far indeed with our neighbour. Its physical condition is still largely a mystery, but we know now that even in its equatorial region the mid-day temperature barely approaches that of our coldest winter. Its air is much more attenuated than ours, its oceans have shrunk . . . and huge ice caps gather and melt about each pole and periodically inundate its temperate zones. That last stage of exhaustion, which to us is still incredibly remote, has become a present day problem for the inhabitants of Mars.

The story itself is violent one involving the arrival in England of a race of octopoid Martians who encase themselves inside tripod-like machines, immune to anything that the army or navy can throw at them. The Martians set about conquering the English countryside with their tripods and a vaporising death ray and, just as it looks as though humanity is doomed, the tripods fall silent, their occupants having been killed by exposure to the Earth's bacteria.

The quality of Wells's writing and the comtinuing promotion of the canals of Mars by Percival Lowell ensured that the book was, and still is, hugely popular. It is still rated as being one of the greatest science fiction works of all time—and deservedly so. However, it also planted an image of the inhabitants of Mars being an envious and aggressive empire-building race with their eyes (or whatever they use to see with) constantly turned towards Earth. This theme was to become severely exploited during the latter half of the twentieth century.

Although Wells had his imitators, the next popular works to deal with Martians were published in 1912 when the creator of *Tarzan*, Edgar Rice Burroughs, contributed two breathtaking series to the periodical *All-Story* concerning the adventures of John Carter, a space explorer. Carter's exploits were exclusively set on Mars, where he encountered many different types of being (or 'Bassooms', as he called them) and had many adventures, each of which, in true periodical style, would end in a cliffhanger, forcing the reader to buy the next issue. These adventures were to eventually be collected together in eleven books, some of which are still in print today. The *John Carter on Mars* series[176–185] is an acknowledged influence on many other science fiction writers, including Ray Bradbury, who was to later write another Mars classic, *Silver Locusts*. In retrospect, it has been argued that Edgar Rice Burroughs was a promoter of the romantic notion about life on Mars whilst H. G. Wells was the pessimist.

Subsequent to Wells and Burroughs, many books were written about Martians but none had quite the same impact as these two authors' works. Instead, the Martians found a far more influential format in which they could live, breed and invade the public's imagination to far greater effect than in print. This format was cinema.

MARTIANS ON FILM

Martians featured little in pre-Second World War films apart from the notable exception *Flash Gordon's Trip to Mars* (1938), which was a Burroughs-style serial in which the Earth had to be saved from the evil intentions of a mad dictator on Mars. Later, some people found analogies between the Flash Gordon serials and the public's obsession with the rise to power of Nazi Germany and the possibility of another war in Europe. Given that these serials were made in the late 1930s, it is certainly a possibility that there would be some reflection of national anxiety about dictatorships, invasion and warfare within them. It is, however, in the post-war cinema that we can find the greatest use of Martians as an expression of a nation's collective paranoia.

The 1950s heralded the birth and growth of teenage culture in the Western world, which brought with it new styles of fashion, music and cinema. The new teenage marketplace demanded that things be brighter, louder and more anti-social than before in a wave of rebellion against established society. In this environment the UFO phenomenon rose to prominence, and this fostered a wave of films involving flying saucers visiting Earth. In this pre-television world teenagers across the Western world would spend their spare time sitting in drive-ins and cinemas to watch the latest offerings from Hollywood.

Science fiction films were a particularly popular genre as they induced controlled feelings of fear amongst the audience, causing the excitement of adrenalin moving about the body and also causing girls to move closer to their dates in anxiety. These films were also cheap to make and easy to write, generally following a formula wherein aliens would invade the Earth, either directly in flying saucers or indirectly by disguising themselves as humans, hell bent on converting human society to their alien way of life. Invading Martians played a key role in many of these films, including classics such as the Oscar-winning adaptation of *The War of the Worlds* (1953) and others such as *Invaders from Mars* (1953).

When these films are viewed now, it can be seen that the invading Martians are a thinly disguised analogy to the Cold War threat from Russia and China perceived by the Western world. Much of the McCarthyite anti-communist hysteria can be seen in the invading Martians, who, after all, did originate from the Red Planet. For example, *The War of the Worlds*, updated in 1953 to be set in small-town America, depicts an unprovoked attack by a race of Martians envious of the riches of the Earth. This invasion is unstoppable by the army, navy or air force, and the government is forced to use its ultimate weapon, the atom bomb, which also has little effect. The invaders keep on marching across the countryside, destroying all in their path, including cities, until they are brought to a halt by exposure to Earthly bacteria. A similar scenario is invoked in *Invaders from Mars*, again set in a small American town, where the invaders capture people, brainwash them and send them out in the community to do their bidding. The themes of these two films, plus the literally tens of other alien invasion films made at that time, encapsulate the two main perceived fears from communism—a direct invasion by an army or the gradual conversion of the general public to the communist cause by propaganda or brainwashing. These films inevitably resulted in either the Americans or British triumphing over the invaders or the Martians themselves winning and bringing civilisation to its knees under their dictatorial occupation. Again, these are clear analogies to the two perceived outcomes of any war with an invading communist nation. Not all films were quite this pessimistic, and there are occasional exceptions, such as the brilliant *The Day the Earth Stood Still* (1951), where an alien comes to warn Earth that its inhabitants' self-destructive nature is threatening the planet and the planets nearby.

By the early 1960s the terms 'Martian', 'Little Green Men' and 'Invaders from Mars' were in such common use that they no longer held the fear that they once did and Martians started to turn up in everything from comic strips to cartoons. Indeed, Marvin the Martian's appearance in the Bugs Bunny cartoons were popular at the time, with Bugs Bunny constantly battling against Marvin's attempts to blow up Earth because it blocked his view of Venus.

As the scientific realisation that Mars was in fact a cold, inhospitable planet slowly became apparent, so the aliens in science fiction films no longer came from Mars but from distant galaxies or unspecified areas of the universe. Apart from the spoof film *Santa Claus Conquers the Martians* (1964), one of the last major depictions of Martians was in a tongue-in-cheek series of bubble gum cards, called *Mars Attacks!*, that were first released in 1962. The focus of these cards was a series of grotesque miniature Martians who had their brains on the outside of their heads and green, skeleton-like frames with a warty green skin. It was the express intention of their designer, Len Brown, for each card to be 'totally frightening so that if a kid bought a pack of five cards there wouldn't be a dull one in there'. In this respect the cards were highly successful and depicted the Martians running amok across the countryside, killing and destroying everything in their path. Some of the scenes were too much for people to handle. In particular, one card depicting a pet dog being blasted by a ray gun whilst a little boy looks on in tears caused such outrage that the cards were withdrawn. In 1996 film director Tim Burton revived the characters on the bubble gum cards in his hugely popular film entitled *Mars Attacks!*, wherein the same bubble-brained Martians invade Earth and destroy society as we know it. The scene involving the shot dog was, however, removed before filming started.

Science fiction became more subtle in its approach to the subject of aliens during the 1960s and 1970s with films and novels such as *2001: A Space Odyssey* (1968), *Barbarella* (1967), *The Planet of the Apes* (1968) and *Star Wars* (1977) portraying complex alien civilisations or fantastic scenarios as opposed to the stark invasion plots of the 1950s. Martians barely receive a mention, although Mars does still occasionally feature.

Ray Bradbury's novel *Silver Locusts* was dramatised as *The Martian Chronicles*, in which human settlers on Mars have fleeting encounters with the remnants of a race of semi-physical, semi-spiritual beings who are wary of their human invaders. In a twist of irony, the world is destroyed by a nuclear explosion, leaving the colonists on Mars as the last surviving humans in the solar system. In a reflection of the scientific attitude to Mars, there is a line in the book where an alien woman turns to her husband to ask about the possibility of life on Earth. 'Of course not,' replies the husband, 'there's far too much oxygen in the atmosphere for life to exist there.'

In an interesting development on the Martian theme, references have been made to finding not beings themselves, but relics of their existence on the surface of Mars. This theme can be found in episodes of *The Twilight*

Zone, The Outer Limits and *The X-Files* and in the film *Total Recall* (1988), and it would seem to be a means of integrating one of the themes of this book, that widespread life may not be present now but that perhaps it was in the past. It also echoes the debate that exists over the 'Face on Mars' theories outlined in Chapter 14.

The notion of the conspiracy theory and government cover-up is also reflected in the 1978 film *Capricorn One*. The plot concerns three astronauts who are waiting to blast off to become the first men on Mars. Just before take off they are kidnapped by the government, who believe that it is cheaper to reconstruct Mars on a filmset and film the astronauts on it rather than fly them there.

The heyday for Martian science fiction was in the 1950s, and today's more educated audiences would not tolerate the thought that beings could live unprotected on the planet. Instead, they now demand complex block-buster films full of special effects and uninhibited violence. There are still the alien invasion films, such as the massively successful *Independence Day* (1996) and *Men in Black* (1997), but there is little room for Martians in to-day's science fiction fraternity.

REFERENCES

Titles annotated with an asterisk (*) are particularly recommended for more in-depth reading.

1 Adams, W., and Dunham, T., 'The B Band of Oxygen in the Spectrum of Mars', *Astrophysical Journal*, Vol. 79 (1934), pp. 308–16.

2 Adams, W., and Dunham, T., 'Water Vapour Lines in the Spectrum of Mars', *Publications of the Astronomical Society of the Pacific*, Vol. 49 (1937), pp. 209–11.

3 Adamski, G., and Leslie, D., *Flying Saucers have Landed!*, Werner Laurie (1953).

4 Agnes, G. (ed. Clerke, A.), *A Popular History of Astronomy during the Nineteenth Century*, Edinburgh (1885).

5 Allingham, C., *Flying Saucers from Mars*, Frederick Muller (1954).

6 Andrewes, C., *The Common Cold*, Cambridge University Press (1965).

7 Antoniadi, E., 'Reports of Mars Section, 1896', *Memoirs of the British Astronomical Association*, Vol. 5 (1897), pp. 82–90.

8 Antoniadi, E., 'Fifth Interim Report for 1909', *British Astronomical Association Journal*, Vol. 20 (1909), pp. 136–41.

9 Antoniadi, E., *La Planète Mars*, Herman (1930).

10 Arrhenius, S., *Destinies of the Stars*, Putnam (1918).

11 Arrhenius, G. and Mojzsis, S., 'Extraterrestrial Life: Life on Mars—Then and Now', *Current Biology*, Vol. 6 (10) (1996), pp. 1213–16.

12 Ash, R., Knott, S., and Turner, G., 'A 4-Gyr Shock Age for a Martian Meteorite and Implications for the Cratering History of Mars', *Nature*, Vol. 380 (1996), pp. 57–9.

13 Avduevsky, V., Akim, E., Aleshin, V., Borodin, N., Kerzhanovich, V., Malkov, V., Marov, M., Morozov, S., Rozhdestvenskiy, M., Ryabov, O., Subbotin, M., Suslov, V., Cheremukhina, Z., and Shkirina, V., 'Martian Atmosphere in the Landing Descent Module of MARS-6 (Preliminary Results)', translated in NASA TT-F-16336 (1975).

14 Bada, J., and McDonald, G., 'Detecting Amino Acids on Mars', *Analytical Chemistry News and Features* (1 November 1996), pp. 668–73A.

15 Barengoltz, J., and Stabekis, P., 'U.S. Planetary Protection Program', *Advances in Space Research*, Vol. 3 (8) (1983), pp. 5–12.

16 Battersby, S., 'Mars Rover Meets Rock with Complex Past', *Nature*, Vol. 388 (1997), p. 215.

17 Becker L., Glavin D., and Bada, J., 'Polycyclic Aromatic Hydrocarbons (PAHs) in Antarctic Martian Meteorites, Carbonaceous Chondrites, and Polar Ice', *Geochimica et Cosmochimica Acta*, Vol. 61 (2) (1997), pp. 475–81.

18 Beer, W, and Madler, J., *Fragmente sur les Corps Celestes du System Solar*, Paris (1840).

19 Bergier, J., *Mysteries of the Earth*, Futura (1973).

20 Berzelius, J., 'Ueber Meteorsteine', *Annalen der Physik und Chemie*, Vol. 33 (1834), pp. 113–48.

21 Bizony, P., *The Rivers of Mars—The Search for the Cosmic Origins of Life*, Aurum Press (1997).

22 Bogard, D., and Johnson, P., 'Martian Gases in an Antarctic Meteorite', *Science*, Vol. 221 (1983), pp. 651–4.

23 Boston, P., Ivanov, M., and McKay, C., 'On the Possibility of Chemosynthetic Ecosystems in Subsurface Habitats on Mars', *Icarus*, Vol. 95 (1992), pp. 300–8.

24 Boyd, R., *General Microbiology*, Times Mirror/Mosby College Publications (1988).

25 Brenner, L., *Spaziergange durch das Himmelszelt*, Leipzig (1898).

26 Briggs, M., 'Organic Constituents of Meteorites', *Nature*, Vol. 191 (1961), pp. 1137–40.

27 Briggs, M., and Kitto, G., 'Complex Organic Micro-structures in the Mokoia Meteorite', *Nature*, Vol. 193 (1962), pp. 1126–7.

28 Buffon, C., 'Partiehypothétique', *Histoire Naturelle*, Vol. 2 (1775), pp. 361–564.

29 Burton, C., 'Canals of Mars', *Astronomical Register*, Vol. 20 (1882), p. 142.

30 Cairns-Smith, G., *Genetic Take-over and the Mineral Origins of Life*, Cambridge University Press (1982).

31 Calvin, M., 'The Chemistry of Life: 3. How Life Originated on Earth and in the World Beyond', *Chemical Engineering News*, Vol. 39 (21) (1961), p. 96.

32 Cameron, R., 'Morphology of Representative Blue-green Algae', *Annals of the New York Academy of Sciences*, Vol. 108 (1963), pp. 412–20.

33 Cameron, R., 'Antarctic Soil Microbial and Ecological Investigations', *Research in the Antarctic* (eds Quam and Porter), American Society for the Advancement of Science (1971), pp. 137–89.

34 Campbell, W., 'The Spectrum of Mars', *Publications of the Astronomical Society of the Pacific*, Vol. 6 (1894), pp. 228–36.

35 Campbell, W., quotation taken from a letter written to G. Hale on 11 May 1908 as reported in Sheehan (1996).

36 Capen, C., 'Martian Yellow Clouds—Past and Future', *Sky and Telescope*, Vol. 41 (1971), pp. 2–4.

37 Carlotto, M., *The Martian Enigmas: A Closer Look*, North Atlantic Books (1991).

38 Carr, M., 'The Role of Lava Erosion in the Formation of Lunar Rilles and Martian Channels', *Icarus*, Vol. 22 (1974), pp. 1–23.

39 Carr, M., 'Martian Channels and Valleys: Their Characteristics, Distribution and Age', *Icarus*, Vol. 48 (1981), pp. 91–117.

40 Casani, J., Interviews in the *Daily Mirror* (UK), 26 January 1976, and *The National Enquirer* (USA), 26 March 1976.

41 Cerulli, V., *Marte nel 1896–97*, Collurania (1898).

42 Chambers, P., *Paranormal People*, Blandford (1998).

43 Chang, S., Tolz, J., Kirschvink, J., and Awramik, S., 'Biogenic Magnetite in Stromatolites. II: Occurrence in Ancient Sedimentary Environments, *Precambrian Research*, Vol. 42 (1989), pp. 305–15.

44 Chown, M., 'Homeward Bound', *New Scientist*, Issue 2098 (1997), p. 7.

45 Clark, B., and van Hart, D., 'The Salts of Mars', *Icarus*, Vol. 45 (1981), pp. 70–378.

46 Claus, G., and Nagy, B., 'A Microbiological Examination of Some Carbonaceous Chondrites', *Nature*, Vol. 161 (1961), pp. 594–6.

47 Cloud, P., *Oasis in Space: Earth History from the Beginning*, Norton (1989).

48 Cohen, P., 'Rocking All Over Mars', *New Scientist*, Issue 2091 (1997), p. 6.

49 Coblentz, W., and Lampland, C., 'Further Radiometric Measurements and Temperature Estimates for the Planet Mars', *Scientific Papers of Natural Bur. Studies*, Vol. 22 (1927), pp. 237–76.

50 Copernicus, N., *Commentariolus* (1593). Reprinted and translated in *Three Copernican Treatises*, Dover (1959).

51 Crew, E., 'The Martian Planetarium', *Fortean Times*, Issue 65 (1992), p. 35.

52 Cros, C., *Oeuvres Completes* (1869). Reprinted 1964.

53 Cutts, J., and Blasius, K., 'The Origin of Martian Outflow Channels: The Eolian Hypothesis', *Journal of Geophysical Research*, Vol. 86 (1981), pp. 5075–102.

54 Daniken, E. von, *Chariots of the Gods?*, Souvenir Press (1969).

55 Deamer, D., 'Prebiotic Conditions and the First Cells', in *Fossil Prokaryotes and Protists* (ed. J. Lipps), Blackwell Scientific Publications (1993).

56 Degens, E., and Bajor, M., 'Amino Acids and Sugars in the Brunderheim and Murray Meteorites', *Naturwissenschaffen*, Vol. 39 (1962), pp. 605–6.

57 Dick, T., *The Works of Thomas Dick*, Hartford (1844).

58 Dick, S., *Plurality of Worlds: The Origins of the Extraterrestrial Life Debate from Democritus to Kant*, Cambridge University Press (1982).

59 DiPietro, A., Molenaar, J., and Brandenburg, G., *Unusual Mars Surface Features*, Mars Research (1982).

60 Dollfus, A., 'Mesure de la Quantité de Vapeur d'Eau Contenue dans l'Atmosphère de la Planète Mars', *Comptes Rendu de la Academie Sciences*, Vol. 256 (1963), pp. 3009–11.

61 Dreibus, G., and Wanke, H., 'Mars, a Volatile Rich Planet', *Meteoritics*, Vol. 20 (1985), pp. 367–81.

62 Editorial, *British Astronomical Association Journal*, Vol. 13 (1903) p. 338.

63 Ellsworth, S., in *Almanac for 1785* (ed. Stowell) (1785).

64 Evans, H., 'The Italian Martians', *Fortean Times*, Issue 67 (1993), p. 43.

65 Fellows, O., and Milliken, S., *Buffon*, New York (1972).

66 Fitch, F. and Anders, E., 'Organised Element: Possible Identification in Orgueil Meteorite', *Science*, Vol. 140 (1963), pp. 1097–100.

67 Fitch, F., Schwarcz, H., and Anders, E., '"Organised Elements" in Carbonaceous Chondrites', *Nature*, Vol. 193 (1962), pp. 1123–5.

68 Flammarion, C., *La Planète Mars et ses Conditions d'Habitabilité*, Vol. 1, Gauthier-Villars et Fils (1892). *Probably still the best summary of the pre-twentieth century history of the search for life on Mars.

69 Flammarion, C., *La Planète Mars et ses Conditions d'Habitabilité*, Vol. 2, Gauthier-Villars et Fils (1909). *An extremely good, if somewhat biased, account of the canals issue.

70 Flournoy, T., *From India to the Planet Mars*, Harper and Bros (1900).

71 Fontenelle, B. de, *Entretiens sur la Pluralité des Mondes* (1686). Translated and reprinted as *Conversations on the Plurality of Worlds* by H. Hargreaves, University of California Press (1990).

72 *Fortean Times*, 'Martian Panics', Issue 43 (1985), p. 12.

73 Fortey, R., *Life: An Unauthorised Biography*, Harper-Collins (1997).

74 Friedmann, E., and Ocampo, R., 'Endolithic Blue-green Algae in the Dry Valleys: Primary Producers in the Antarctic Desert Ecosystem', *Science*, Vol. 193 (1976), pp. 1247–9.

75 Galton, F., letter to *The Times* (London), 6 August 1892.

76 Gifford, F., 'The Martian Canals According to a Purely Aeolian Hypothesis', *Icarus*, Vol. 3 (1964), pp. 130–5.

77 Grady, M.,Wright, I., Douglas, C., and Pillinger, C., 'Carbon and nitrogen in ALH84001', *Meteoritics*, Vol. 29 (1994), p. 469.

78 Green, N., quotation in *Astronomical Register*, Vol. 16 (May 1878).

79 Green, N., 'The Canals of Mars', *Astronomical Register*, Vol. 18 (1880), p. 138.

80 Hansson, A., *Mars and the Development of Life*, 2nd edn, John Wiley and Sons/Praxis Publishing (1997). *An analysis of the evolution of life and its application to Mars, written in the style of a reference book.

81 Harvey, R., and McSween, H., 'A Possible High-Temperature Origin for the Carbonates in the Martian Meteorite ALH84001', *Nature*, Vol. 382 (1996), pp. 49–51.

82 Haweis, K., quoted in 'The Opposition of Mars' by J. Lockyer, *Nature*, Vol. 46 (1892), pp. 443–8.

83 Hayatsu, R., 'Orgeuil Meteorite: Organic Nitrogen Contents', *Science*, Vol. 146 (1964), pp. 1291–2.

84 Herschel, J., *Outlines*, London (1858).

85 Herschel, W., 'On the remarkable Appearances at the Polar Regions of the Planet Mars, the Inclination of its Axis, the Position of its Poles, and its spheroidical Figure; with a few hints relating to its real Diameter and Atmosphere', *The Philosophical Transactions of the Royal Society of London*,

Vol. 74 (1784), pp. 233–73.

86 Hess, S., Henry, R., Leovy, C., Ryan, J., and Tillman, J., 'Meteorological Results from the Surface of Mars: Viking 1 and 2', *Journal of Geophysical Research*, Vol. 82 (1977), pp. 4559–74.

87 Hoagland, R., *The Monuments of Mars: A City on the Edge of Forever*, North Atlantic Books (1987). *This book is the one that sparked widespread interest in the Face on Mars.

88 Hochstein, L., and Morowitz, H., *Workshop on the Occurrence of Halophilic Bacteria in Bedded Salt Deposits*, Fairfax, Va (1994).

89 Holland, H., 'Evidence for Life on Earth More Than 3850 Million Years Ago', *Nature*, Vol. 275, Pt 3 (1997), pp. 38–9.

90 Hope, E., *The Question of the Tectonic Origin of Linear Formations on Mars*, Report No T 445R (1966), Defence Research Board, Canada.

91 Horneck, G., 'Responses of *Bacillus subtilis* Spores to Space Environment—Results from Experiments in Space', *Origins of Life and Evolution of the Biosphere*, Vol. 23 (1) (1993), pp. 37–52.

92 Horneck, G., 'Long Term Survival of Bacterial Spores in Space', *Advances in Space Research*, Vol. 14 (10) (1994), pp. 41–5.

93 Horowitz, N., 'Life on Mars—A Reply', *Origins of Life and Evolution of the Biosphere*, Vol. 18 (1988), pp. 309–10.

94 Hoyle, F., and Wickramasinghe, C., *Evolution from Space*, J. W. Dent (1981).

95 Hoyle, F., and Wickramasinghe, C., *Our Place in the Cosmos*, J. W. Dent (1993).

96 Hoyle, F., and Wickramasinghe, C., *Life on Mars?*, Clinical Press (1997).

97 Hubbard, J., Hardy, J., and Horowitz, N., 'Photocatalytic Production of Organic Compounds from CO and H_2O in a Simulated Martian Atmosphere', *Proceedings of the National Academy of Science*, Vol. 68 (1971), pp. 574–8.

98 Huber, R., Stoffers, P., Cheminee, J., Richnow, H., and Stetter, K., 'Hyperthemophilic Archaebacteria within the Crater and Open-sea Plume of Erupting Macdonald Seamount', *Nature*, Vol. 345 (1990), pp. 179–81.

99 Huguenin, R., Miller, K., and Leschine, S., 'Mars: A Contamination Potential?', *Advances in Space Research*, Vol. 3 (8) (1983), pp. 35–8.

100 Hutchins, K., and Jakosky, B., 'Carbonates in Martian Meteorite ALH84001: A Planetary Perspective on Formation Temperature', *Geophysical Research Letters*, Vol. 24 (7) (1997), pp. 19–822.

101 Huygens, C., *The Celestial Worlds Discover'd or, Conjectures concerning the Inhabitants, Plants and Productions of the Worlds in the Planets*, London (1698).

102 Hyslop, J., *Psychical Research and the Resurrection*, Small/Maynard (1908).

103 *Independent* (periodical) 'Communicating with Mars', (1909), pp. 1042–3.

104 Inglis, B., *The Natural and Supernatural*, 2nd edn, Prism Press (1992).

105 Jagoutz, E., Sorowka, A., Vogel, J., and Wanke, H. 'ALH84001: Alien or

Progenitor of the SNC Family?', *Meteoritics*, Vol. 29 (1994), pp. 478–9.

106 Joly, J. 'On the Origin of the Canals of Mars', *Scientific Transactions of the Royal Dublin Society*, Series 2, Vol. 6 (1898), pp. 249–68.

107 Jull, A., Eastoe, C., Xue, S., and Herzog, G., 'Isotopic Composition of Carbonates in the SNC Meteorites Allan Hills-84001 and Nakhla', *Meteoritics*, Vol. 30 (1995), pp. 311–18.

108 Jull, A., Eastoe, C. and Cloudt, S., 'Isotopic Composition of Carbonates in the SNC Meteorites, Allan Hills 84001 and Zagami', *Journal of Geophysical Research*, Vol. 102 (E1) (1997), pp. 1663–9.

109 Jung, C., *Zur Psychologie und Pathologie sogenannter Occulter Phanomene*, Muntze (1902).

110 Kaplan, L., Munch, G., and Spinard, H., 'An Analysis of the Spectrum of Mars', *Astrophysical Journal*, Vol. 139 (1964), pp. 1–15.

111 Kappen, L., 'Response to Extreme Environments', in *The Lichens* (eds Ahmadjian and Hale), Academic Press (1973), pp. 311-380.

112 Katterfel'd, G., *The Question of the Tectonic Origin of Linear Formations on Mars*, Izv. Vses. Geogr. Obschch (1959).

113 Kepler, J., *Astronomia Nova* (1605). Reprinted (trans. W. Donohue) as *New Astronomy*, Cambridge University Press (1993).

114 Kerr, R., 'Pathfinder Strikes a Rocky Bonanza', *Science*, Vol. 277 (1997), pp. 173–4.

115 Kieffer, H., Jakosky, B., Snyder, C., and Matthews, M., *Mars*, Arizona University Press (1992). *Undoubtedly the best summary of the scientific knowledge of Mars before the arrival of Pathfinder.

116 Kiernan, V., 'NASA Told How to Avoid Mars Attacks', *New Scientist*, Issue 2074 (1997), p. 6.

117 Kirschvink, J., Maine, A., and Vali, H., 'Paleomagnetic Evidence of a Low-Temperature Origin of Carbonate in the Martian Meteorite ALH84001, *Science*, Vol. 275 (5306) (1997), pp. 1629–33.

118 Klein, H., 'The Viking Biological Experiments on Mars', *Icarus*, Vol. 34 (1978), pp. 666–74. *A good summary of the possibility of life on Mars based on the Viking data.

119 Klein, H., Horowitz, N., and Biemann, K., 'The Search for Extant Life on Mars', in *Mars* (eds Kieffer *et al*), Arizona University Press (1992), pp. 1221–33.

120 Kuiper, G., 'Survey of Planetary Atmospheres', *Contributions from the McDonald Observatory*, No 161 (1949), pp. 304–45.

121 Kuiper, G., 'Note of Dr. McLaughlin's Paper', *Publications of the Astronomical Society of the Pacific*, Vol. 68 (1956), pp. 304–5.

122 Kuiper, G., 'Visual Observations of Mars, 1956', *Astrophysical Journal*, Vol. 125 (1957), pp. 307–17.

123 Lake, J., 'Origin of Eukaryotic Nucleus Determined by Rate-invarient Analysis of rDNA Sequences', *Nature*, Vol. 331 (1988), pp. 184–6.

124 Leshin, L., Epstein, S., and Stolper, E., 'Hydrogen Isotope Geochemistry of SNC Meteorites', *Geochimica et Cosmochimica Acta*, Vol. 60 (14) (1996),

pp. 2635–50.

125 Levin, G., and Stratt, P., 'Antarctic Soil No. 726 and Implications for the Viking Labelled Release Experiment', *Journal of Theoretical Biology*, Vol. 91 (1981), pp. 41–5.

126 Levin, G., and Stratt, P., untitled, in 'The NASA Mars Conference' (ed. Reiber), *American Astronautical Society* (1986), pp. 187–208.

127 Levinthal, E., Jones, K., Fox, P., and Sagan, C., 'Lander Imaging as a Detector of Life on Mars', *Journal of Geophysical Research*, Vol. 82 (28) (1977), pp. 4468–78.

128 Lowell, P., *Mars*, Houghton Mifflin (1895).

129 Lowell, P., interview in *Boston Evening Telegraph*, 28 November 1896.

130 Lowell, P., *Mars and its Canals*, Macmillan (1906).

131 Lowell, P., *Mars as the Abode for Life*, Macmillan (1910). *All of Lowell's theories collected into one Vol..

132 Lucchitta, B., 'Valles Marineris—Faults, Volcanic Rocks, Channels, Basin Beds', in *Reports of the Planetary Geology Program*, NASA TM 84211 (1981), pp. 419–21.

133 Lund, J., 'Soil Algae', in *Physiology and Biochemistry of Algae* (ed. Lewin), Academic Press (1962), pp. 759–70.

134 MacDermott, A., Barron, L., Brack, A., Buhse, T., Drake, A., Emery, R., Gottarelli, G., Greenberg, J., Haberle, R., Hegstrom, R., Hobbs, K., Kondepudi, D., McKay, C., Moorbath, S., Raulin, F., Sandford, M., Schwartzman, D., Thiemann, W., Tranter, G., and Zarnecki, J., 'Homochirality as the Signature for Life: The SETH Cigar', *Planetary Space Science*, Vol. 44 (11) (1996), pp. 1441–6.

135 Maher, K., and Stevenson, D., 'Impact Frustration of the Origin of Life', *Nature*, Vol. 331 (18 February 1988), pp. 612–14.

136 Matthews, R., and MacKenzie, D., 'Earthlings Wrangle over Search for Martians', *New Scientist*, Issue 2086 (1997), p. 12.

137 Maunder, E., 'The Canals of Mars', *Knowledge* (1 November 1894), pp. 249–52.

138 Maunder, E., and Evans, J., 'Experiments as to the Actuality of the "Canals of Mars"', *Monthly Notices of the Royal Astronomical Society*, Vol. 58 (1903), pp. 488–99.

139 McDaniels, S., *The McDaniel Report*, North Atlantic Books (1993).

140 McLaughlin, D., 'Volcanism and Aeolian Deposition on Mars', *Geological Society of America Bulletin*, Vol. 65 (1954), pp. 715–17.

141 McKay, C., 'Looking for Life on Mars', *Astronomy*, August 1997, pp. 38–43.

142 McKay, C., and Nedell, S., 'Are There Carbonate Deposits in the Valles Marineris, Mars?', *Icarus*, Vol. 73 (1988), pp. 142–8.

143 McKay, C., Mancinelli, R., Stoker, C., and Wharton, R. 'The Possibility of Life on Mars During a Water Rich Past', in *Mars* (eds Kieffer *et al*), University of Arizona Press (1992), pp. 1234–48.

144 McKay, D., Gibson, E., Thomas-Keprta, K., Vali, H., Romanek, C.,

Clemett, S., Chiller, X., Maechling, C., and Zare, R., 'Search for Past Life on Mars: Possible Relic Biogenic Activity in Martian Meteorite ALH84001', *Science*, Vol. 273 (1996), pp. 924–30. *The paper that started the ALH 84001 issue, somewhat technical but still very useful.

145 McSween, H., 'What We Have Learned about Mars from SNC Meteorites', *Meteoritics*, Vol. 29 (1994), pp. 757–79.

146 McSween, H., Stropler, E., Taylor, L., Muntean, R., O'Kelley, G., Eldridge, J., Biswas, S., Ngo, H., and Lipschutz, M., 'Petrogenetic Relationship between Allan Hills 77005 and Other Chondrites', *Earth Planet Science Letters*, Vol. 45 (1979), pp. 275–84.

147 Melosh, H., 'The Rocky Road to Panspermia', *Nature*, Vol. 332 (6166) (1988), pp. 687–8.

148 Meinschein, W., Nagy, B., and Hennessy, D., 'Evidence in Meteorites of Former Life: The Organic Compounds in Carbonaceous Chondrites are Similar to those Found in Marine Sediments', *Annals of the New York Academy of Sciences*, Vol. 108 (1963), pp. 553–79.

149 Mittlefehldt, D., 'ALH84001, a Cumulate Orthopyroxenite Member of the Martian Meteorite Clan', *Meteoritics*, Vol. 29 (1994), pp. 214–21.

150 Molesworth, P., quotation found in Sheehan (1996) referenced as taken from a manuscript in the Royal Astronomical Society's archive (1903).

151 Moreno, M., 'Microorganism Transport from Earth to Mars', *Nature*, Vol. 336 (1988), p. 209.

152 Mutch, T., Arvidson, R., Head, J., Jones, K., and Saunders, R., *The Geology of Mars*, Princeton University Press (1976).

153 Myers, D., and Percy, D., *Two-Thirds*, Aulis Press (1994).

154 Nedell, S., Squyres, S., and Andersen, D., 'Origin and Evolution of the Layered Deposits in the Valles Mariners, Mars', *Icarus*, Vol. 70 (1987), pp. 490–1.

155 *New Scientist*, 'Amber Alien Surprises Lazarus Team', Issue 2082 (1997), p. 7.

156 Norton, C., and Grant, W., 'Survival of Halobacteria within Fluid Inclusions in Salt Crystals', *Journal of Genetic Microbiology*, Vol. 134 (1988), pp. 1365–73.

157 O'Leary, M., 'Carbon Isotope Fractionation in Plants', *Phytochemistry*, Vol. 20 (1981), pp. 553–67.

158 Oncley, P., and Fulmer, C., 'The Martian "Canali" as Meteoritic Crater Rays', *Transactions of the American Geophysical Union*, Vol. 47 (1966), p. 482.

159 Opik, E., 'The Surface of Mars', *Science*, Vol. 153 (1966), pp. 255–66.

160 Oro, J., 'Comets and the Formation of Biochemical Compounds on the Primitive Earth', *Nature*, Vol. 190 (4774) (1961), pp. 389–90.

161 Paige, D., 'The Thermal Stability of Martian Ground Ice', *Nature*, Vol. 356 (1992), pp. 43–5.

162 *The Paranormal World of Paul McKenna*, a popular programme shown on British Independent Television during June 1997.

163 Pettit, E., and Richardson, R., 'Observations of Mars made at Mount Wilson in 1954', *Publications of the Astronomical Society of the Pacific*, Vol. 67 (1955), pp. 62–73.

164 Pickering, E., 'Mars', *Astronomy and Astro-physics*, Vol. 2 (1892), pp. 668–72.

165 Pickering, E., quoted in 'Sur les Canaux de la Planète Mars', *Comptes Rendus de la Académie de France* (Frizeau, H.) (1888), pp. 1759–62.

166 Pickering, E., *Mars*, Badger (1921).

167 Pittendrigh, C., Vischniac, W., and Pearman, J., 'Biology and the Exploration of Mars', *National Academy of Science Publication No 1296* (1966).

168 Plassmann, J., *Ist Mars ein bewohnter Planet?*, Frankfurt (1901).

169 Plumb, R., 'Non-problematic Chemistry of Mars and the LR Experiment', Presented at the Ames Conference on Simulation of Mars Surface Properties (1977).

170 Pozos, R., *The Face on Mars: Evidence of a Lost Civilisation?*, Chicago Review Press (1986).

171 Proctor, R., 'Maps and Views of Mars', *Scientific American*, Special Supplement No 26 (1888).

172 Rao, D., and Le Blanc, F., 'Effects of Sulphur Dioxide on the Lichen Algae, with Special Reference to Chlorophyll', *Bryologist*, Vol. 69 (1966), pp. 69–75.

173 Raup, D., and Stanley, S. *Principles of Paleontology*, 2nd edn (1978), Freeman and Co.

174 Rea, D., O'Leary, B., and Swinton, W., 'The Origin of the 3.58 and 3.69 Micron Minima in the Infrared Spectra', *Science*, Vol. 147 (1965), pp. 1286–8.

175 Reiser, R. and Tasch, P., 'Investigations of the Viability of Osmophile Bacteria of Great Geological Age'. *Transactions of the Kansas Academy of Sciences*, Vol. 63 (1960), pp. 31–4.

176 Rice-Burroughs, E., *A Princess of Mars*, London (1912).

177 Rice-Burroughs, E., *The Gods of Mars*, London (1918).

178 Rice-Burroughs, E., *The Warlord of Mars*, London (1919).

179 Rice-Burroughs, E., *Thuvia, Maid of Mars*, London (1920).

180 Rice-Burroughs, E., *The Chessman of Mars*, London (1922).

181 Rice-Burroughs, E., *The Mastermind of Mars*, London (1928).

182 Rice-Burroughs, E., *A Fighting Man of Mars*, London (1931).

183 Rice-Burroughs, E., *Swords of Mars*, London (1936).

184 Rice-Burroughs, E., *Synthetic Men of Mars*, London (1940).

185 Rice-Burroughs, E., *Llana of Gathol*, London (1948).

186 Richardson, R., and Bonestell, C., *Mars*, Allen & Unwin (1965).

187 Rickard, R., and Sievking, P., 'Martian Madness', *Fortean Times*, Issue 103 (1997), p. 4.

188 Robbins, E., and Iberall, A. 'Mineral Remains of Early Life on Earth? On Mars?', *Geomicrobiology Journal*, Vol. 9 (1991). pp. 51–66.

189 Rochas, A. de, *Les Vies Successives*, Charcornac Frères (1924).

190 Romanek, C., Grady, M., Wright, I., Mittlefehldt, D., Socki, R., Pillinger, C., and Gibson, E., 'Hydrothermal Activity on Mars: The Record from ALH84001', *Nature*, Vol. 372 (1994), pp. 655–7.

191 Rothschild, L., 'Earth Analogs for Martian Life. Microbes in Evaporites, a New Model System for Life on Mars', *Icarus*, Vol. 88 (1990), pp. 246–60. *A good discussion as to where life on Mars could currently survive.

192 Rothschild, L., 'A Cryptic Microbial Mat—A New Model Ecosystem for Extant Life on Mars', *Advances in Space Research*, Vol. 15 (3) (1994), pp. 223–8.

193 Sack, R., Azeredo, W., and Lipschutz, M., 'Olivine Diogenites: The Mantle of the Eucrite Parent Body', *Geochima, Cosmochima, Acta*, Vol. 55 (1991), pp. 1111–20.

194 Saeki, S., article in the *Los Angeles Times*, 28 January 1950.

195 Sagan, C., 'High Resolution Planetary Photography and the Detection of Life', *Proceedings of Caltech-JPL Lunar and Planetary Colloq.*, Memorandum 33 (1966), pp. 279–87.

196 Sagan, C., 'The Long Winter Model of Martian Biology: A Speculation', *Icarus*, Vol. 15 (1971), p. 511.

197 Sagan, C., Veverka, J., Fox, P., Bubisch, R., Lederberg, J., Levinthal, E., Quam, L., Tucker, R., Pollack, J., and Smith, B., 'Variable Features on Mars: Preliminary Mariner 9 Television Results', *Icarus*, Vol. 17 (1972), pp. 346–72.

198 Sagan, C., Veverka, J., Fox, P., Bubisch, R., French, R., Gierasch, P., Quam, L., Lederberg, L., Levinthal, E., Eross, B., Tucker, R., and Pollack, J., 'Variable Features on Mars 2: Mariner 9 Global Results', *Journal of Geophysical Research*, Vol. 78 (1973), pp. 4163–96.

199 Sagan, C., and Fox, P., 'The Canals of Mars: An Assessment after Mariner 9', *Icarus*, Vol. 5 (1975), pp. 602–12. *A thorough demolition of the canals using the Mariner photographs.

200 Sagan, C., and Wallace, D. 'A Search for Life on Earth at 100 metre Resolution', *Icarus*, Vol. 15 (1971), pp. 515–54.

201 Sagan, C., Toon, O., and Gierasch, P., 'Climate Change on Mars', *Science*, Vol. 181 (1973), pp. 1045–9.

202 Sargent, K., and Fliermans, C., 'Geology and Hydrology of the Deep Subsurface Microbiology Sampling Sites at the Savannah River Plant, South Carolina', *Geomicrobiology Journal*, Vol. 7 (1989), pp. 3–13.

203 Schiaparelli, G., 'Osservazioni astronomiche e fisiche sull' asse di rotazione e sulla topografia del pianete Marte', *Atti della Royale Academia dei Lincei, Memoria della el di science fisiche*, Vol. 1 (series 3) (1877), pp. 308–439.

204 Schiaparelli, G., 'Osservazioni sulla topografia del pianeta Marte durante l'opposizione 1881–82—Communicazione Preliminare', *Opere*, Vol. 1 (1882), pp. 381–8.

205 Schiaparelli, G., 'The Planet Mars', *Natura ed Arte* (1893), trans. E. Pickering in *Astronomy and Astrophysics*, Vol. 13 (1894), pp. 635–40.

206 Schiaparelli, G., Struve, O., Terby, F., *et al*, *Corrispondenza su Marte*, Domus Galilaeana (1976). A collection of letters written between astronomers, collected into two volumes.

207 Schidlowski, M., 'A 3,800-million-year Isotopic Record of Life from Carbon in Sedimentary Rocks', *Nature*, Vol. 333 (1988), pp. 313–18.

208 Schonfeld, E., 'Martian Volcanism', *Lunar Science VIII*, Lunar Planetary Institute (1977), pp. 843–5.

209 Schopf, J., and Packer, B., 'Early Archean (3.3 to 3.5 Billion Year Old) Microfossils from Warrawoona Group, Australia', *Science*, Vol. 237 (1987), pp. 70–2.

210 Schumm, S., 'Structural Origin of Large Martian Channels', *Icarus*, Vol. 22 (1974), pp. 371–84.

211 *Science News*, 'Facing Up to Mars', 7 August 1976.

212 Scott, D., and Chapman, M., 'Mars Elysium Basin: Geologic/Volumetric Analysis of a Young Lake and Exobiologic Implications', *Proceedings of Lunar and Planetary Science Conference*, Vol. 21 (1991), pp. 669–77.

213 Scott E., Yamaguchi A., and Krot, A., 'Petrological Evidence for Shock Melting of Carbonates in the Martian Meteorite ALH84001', *Nature*, Vol. 387 (1997), pp. 377–9.

214 Secchi, A., *Osservazioni del pianeta Marte. Memorie dell'Osservatorio del Collegio Romano*, Rome (1863).

215 Sharp, R., and Williams, R. 'Properties of *Thermus ruber* Strains Isolated from Icelandic Hot Springs and DNA: DNA Homology of *Thermus ruber* and *Thermus aquaticus*', *Applied Environmental Microbiology*, Vol. 54 (1988), pp. 2049–53.

216 Sheehan, W., *The Planet Mars—A History of Observation and Discovery*, Arizona University Press (1996). *A good summary of the history of the scientific study of Mars.

217 Siegel, S., and Daly, O., 'Responses of *Cladonia rangiferina* to Experimental Climatic Factors', *Botany Gazette*, Vol. 129 (1968), pp. 339–45.

218 Sleep, N., Zahnle, J., Kasting, J., and Mororwitz, H., 'Annihilation of Ecosystems by Large Asteroid Impacts on the Early Earth', *Nature*, Vol. 342 (1989), pp. 139–42.

219 Slipher, E., *The Photographic Story of Mars*, Sky Publication Corp. (1962).

220 Spence, L., *The Encyclopaedia of the Occult*, Bracken Books (1994; reprinted from the first 1920 edition).

221 Spinard, H., Munch, G., and Kaplan, L., 'The Detection of Water Vapour on Mars', *Astrophysical Journal*, Vol. 137 (1963), pp. 1319–21.

222 Squyres, S., and Kastings, J., 'Early Mars: How Warm and Wet?', *Science*, Vol. 262 (1994), pp. 744–9.

223 Squyres, S., Clifford, S., Kuzmin, R., Zimbelman, J., and Costard, F., 'Ice in the Martian Regolith', in *Mars* (eds Kieffer *et al*), University of Arizona Press, pp. 523–56.

224 Swift, J., *Gulliver's Travels*, London (1726; reprinted many times since).

225 Temple, R., *The Sirius Mystery*, Futura (1976).

226 Tesla, N., 'Talking with the Planets', *Collier's Weekly*, Vol. 24 (1901), pp. 4–5.

227 Thomas, D., and Schimel, J., 'Mars After the Viking Missions: Is Life Still Possible?', *Icarus*, Vol. 91 (1991), pp. 199–206.

228 Thorpe, T., 'Viking Orbiter Observations of the Mars Opposition Effect', *Icarus*, Vol. 36 (1978), pp. 204–15.

229 Tolla, L. de, 'Past Life on Mars?', *Science*, Vol. 273, letters (20 September 1996).

230 Tombaugh, C., 'Geological Interpretation of the Markings on Mars', *Astronomy Journal*, Vol. 55 (1950), p. 184.

231 Trench, S., *The Flying Saucer Story*, Tandem (1966).

232 Tsiolkovskii, K., trans. as 'Can the Earth ever inform the inhabitants of other planets about the existence of intelligent beings on it?, in *Interplanetary Flight and Communication*, Vol. 1 (3) (1971), pp. 53–5.

233 Urey, H., 'Life-forms in Meteorites. Origin of Life-like Forms in Carbonaceous Chondrites', *Nature*, Vol. 193 (1962), pp. 1119–33.

234 Urey, H., 'The Origin of Some Meteorites from the Moon', *Naturwissenschaffen*, Vol. 55 (1963), pp. 49–55.

235 Valley, J., Eiler, J., Graham, C., Gibson, E., Romanek, C., and Stolper, E., 'Low-temperature Carbonate Concretions in the Martian Meteorite ALH84001: Evidence from Stable Isotopes and Mineralogy', *Science*, Vol. 275 (5306) (1997), pp. 1633–8.

236 Vaucouleurs, G. de, (trans. Moore, Patrick), *Physics of the Planet Mars*, Faber & Faber (1954).

237 Vishniac, W., Atwood, R., Bock, R., Gaffron, T., Jukes, T., McLaren, A., Sagan, C., and Spinrad, H., 'A Model of Martian Ecology', in *Biology and the Exploration of Mars* (eds Pittendrigh, Vishniac and Pearman), National Academy of Sciences Publication No 1296 (1966), pp. 229–42.

238 Wadwha, M., and Lugmair, G., untitled, *Meteoritics and Planetary Science*, Vol. 31 (1996), p. 145.

239 Wallace, A., *Is Mars Habitable?*, Macmillan (1907).

240 Wasiutynski, J., 'Studies in the Hydrodynamics and Structure of Stars and Planets', *Astrophysics Norvegica*, Vol. 4 (1946), pp. 241–4.

241 Wells, H., *The War of the Worlds*, Harper (1898).

242 Wharton, R., Crosby, J., McKay, C. and Rice, J., 'Paleolakes on Mars', *Journal of Palaeolimnology*, Vol. 13 (1995), pp. 267–83.

243 Wilson, C., *Mysteries*, Granada (1978).

244 Wilson, C., *The Psychic Detectives*, Pan (1984).

245 Woese, C., 'Bacterial Evolution', *Microbiology Review*, Vol. 51 (1987), pp. 221–71.

246 Wohler, M., and Hornes, M., 'Meteorites', *Sitzungsberichte Akademie der Wissenschaften Wien, Math-Naturw*, Vol. 34 (1859), p. 7.

247 Wordsworth, W., quoted from *Religous Trends in English Poetry*, Vol. 1 (ed. H. Fairchild), New York (1931).

248 Wright, I., Grady, M., and Pillinger, C., 'Organic Materials in a Martian Meteorite', *Nature*, Vol. 340 (1989), pp. 220–2.

INDEX